Empty your mind of all thoughts.
Let your heart be at peace.
Watch the turmoil of beings,
but contemplate their return.
Each separate being in the Universe
returns to the common source.
Returning to the source is serenity.
If you don't realize the source,
you stumble in confusion and sorrow.
When you realize where you come from,
you naturally become tolerant,
disinterested, amused,
kindhearted as a grandmother,
dignified as a king.
Immersed in the wonder of the Tao,
you can deal with whatever life brings you,
and when death comes, you are ready.

– Lao Tzu

Evolution Publications
Sacramento, California 95822

The medical and health procedures in this book are based on training, personal experience, and research of the author. Because each person and situation is unique, the author and publisher urge the reader to research themselves before using any procedure where there is any question as to its appropriateness.

Because there is some risk involved, the author and publisher are not responsible for any adverse effects or consequences resulting from the use of any of the suggestions, preparations, or procedures in this book. Feel free to consult a physician or other qualified health professional.

Cover Art by J. Jahi Bentley
Cover Design by J. Jahi Bentley

ISBN 13: 978-0-615-41591-8
ISBN 10: 0-615-41591-1

Copyright © 2010 by J. Jahi Bentley
http://www.jameljahi.com

All rights reserved. No part of this publication may be reproduced, stored in a retrieval system, or transmitted, in any form or by any means, electronic, mechanical, photocopying, recording, or otherwise, without the prior written permission of the copyright owner.

Printed in the United States of America

How could I ever thank my mother for life? The love, compassion, insight, and wisdom she gave helped me develop my own voice and a powerful source of inner strength. The values she instilled in me empowered me to strive to be a supreme being. Mom, you're a Goddess a Empress!!!

Thank you for the souls that have assisted me throughout my personal evolution and during this journey of life. I am humbled... Each person that has crossed my path has given me wisdom and insight into myself; I & I. There is no me;
Thou Art That.....

The Illuminism of Reality

By

J. Jahi Bentley

MOORISH-AMERICAN EDITION

Printed in the United States of America

EVOLUTION PUBLICATIONS
SACRAMENTO, U. S. A.

Part 1: Knowledge

Chapter 1 - The Mind............Pg 19

Chapter 2 - Emotions...........Pg 28

Chapter 3 - Religion...........Pg 35

Chapter 4 - History............Pg 64

Chapter 5 - Society............Pg 85

Chapter 6 - The Universe.......Pg 104

Part 2: Wisdom

Chapter 7 - Yin and Yang.......Pg 114

Chapter 8 - Treasures of Life...Pg 118

Chapter 9 - Human Organism.....Pg 127

Chapter 10 - Five Elements......Pg 134

Chapter 11 - Twelve Meridians...Pg 139

Chapter 12 - Water..............Pg 144

Chapter 13 - Nutrition..........Pg 149

Part 3: Overstanding

Chapter 14 - Yoga & Meditation...Pg 166

Chapter 15 - The Chakras........Pg 190

Chapter 16 - Pranayama..........Pg 201

Chapter 17 - Quantum Reality....Pg 210

Preface

In the midst of the boundless universe, I arrived on earth physically as a fetus resting within the embryonic sac of my mother's womb; a student in this human experience, a scientist, an artist, a giver, a taker, and a friend. Winning, failing, learning, and creating; often flawed and often magnificent. The precious thoughts of superior minded men and woman have guided me from the wilderness of ignorance and uncertainty to a state of clarity and personal evolution; developing a consciousness submerged in truth, wisdom, spiritual insight, and the reality of immortality. As an aspirant for higher knowledge, I have learned in order to understand the causes which have produced the effects we experience today, history must be studied. You must be strong enough and serine enough to see the historical facts and to interpret them. We can all be impassioned but this is not how we will resolve the very complex problems of humanity; this is why the study of history gives us the serenity required to appreciate the facts as they are. The condition of existence with a physical form is seen as an aberration from the original essence. Soul is our true essence; the original state of infinite consciousness, immortality, and energy. Consciousness is the underlying support of all things in the universe. Matter cannot exist without consciousness to give it form and to be the perceiver of its existence, because matter is only an illusion projected by the conscious perceiver that uses sensory organs to perceive with and a mind to interpret that which is perceived. The world and everything in it is only a reflection of the absolute reality. What we seem to perceive with our senses is in reality, only different aspects of the same substance. Gross matter emanates from and is sustained by the intelligent cosmic vibration, the subtle building material of the universe. Therefore, neither the body nor the senses constitute the essence of what we call consciousness. Our senses contribute to our ignorant concept of reality by supplying the mind with limited information as to the nature of reality. When the unconditioned consciousness enters the body and interacts with the world through the perception instruments of the senses it forgets its unconditioned state, and mental complexes originate in the attachment of the soul to the physical illusion of reality culminating with the ultimate mental complex: the identification with the physical form or body, which is the ego. The mind is the cause of this bondage and release, which discerns and discriminates between various states of consciousness. This discrimination is an apparent duality whereby the one reality (energy or consciousness), seems to be many realities. Creation occurs due to the power of the mind to think and know in what is thought. It is because of the veil of ignorance as to the true nature of the soul that we interact with the universe as if it really exists as an entity other than ourselves. The universe proceeds from the mind of the creator. All things, our mental ideals as well as physical reality are in reality emanations from the cosmic mind of the creator (energy or consciousness). Therefore, the alignment of the individual human mind with the cosmic mind will bring forth union with the cosmos. The unconditioned mind is the source of release or freedom from the fetters that hold the self chained to the events of the world, and our perceptions of them. Thus, by getting back to the source of the original thoughts of the mind you become one with the infinite consciousness understanding the unreality of matter. Our true self is our conscience. Our soul gives life to the body, senses, and mind, and therefore everything is a reflection of the degree of ignorance or wisdom, of the true self, which is within our consciousness.

Forward

I think the most important thing to remember is to stay in the present moment. When I am anxious or afraid my mind drifts into past and future, and I might start to feel like I am floating away without any solid foundation or structural security. I'm listening to and trusting my intuition, here right now. It has never been natural to want to take things slow and careful but I've listened to that, embraced it, and I'm flowing with it. I'm being honest with my entire self... including the contradictions. I am human. We live in paradox, and there are many layers and dimensions to us that are difficult to reason with in terms of our linear thinking. Our spoken language is very linear, and riddled with polarities: you/me, us/them, here/there, up/down. Our intuition is non-linear, multi-layered, and multi-dimensional. So while it is most challenging for me to verbally communicate my intuition, I know that it is truly our higher wisdom, our deeper connection to the world, to those we love, and the interconnected consciousness. I've tried to embrace what I feel and be with it while trying to maintain the morals and values instilled in me. My fears, and even my frustrations are still derived by a deeper foundation of love and honesty and this is my center; my deepest truth. So I'm striving to embrace and flow and dance joyfully with these elements of challenge, as there is no use trying to push them away. Life has taught me that challenge is the catalyst of evolution, inspiration. The highest knowledge is unutterable, for it exists as an entity in lanes which transcend all material words or symbols. All symbols are but keys to doors leading to truths, and many times the door is not opened because the keys seem so great that the things which are beyond them are not visible. When I began to understand that all keys, all material symbols are manifestations; extensions of a great law and truth, I began to develop the vision which has enabled me to penetrate beyond the veil. All things in all universes move according to law, and the law which regulates the movement of the planets is no more immutable than the law which regulates the material expressions of the human being.

Introduction

What, if anything, is there that I could possibly say to free your mind? At least be open to listen and completely internalize my thoughts on the page, because for me to recollect the process of my existence verbally in itself has been my deepest challenge as a human being. To be honest, I don't know where to begin. I've spent my entire life thinking and contemplating life, existence, the universe; I mean I've been into it, deeply since I was a child. I guess you could say I was extremely inquisitive early. I had a really happy childhood; I love sightseeing, traveling, shogi and chess. I just love to have fun, and enjoy the air, people, nature, etc. I don't think it was until I was maybe eighteen or so that I truly started to think about knowledge of self, or consciousness as some would say. I think music has played a major role in my evolution as a human being, individual, man, and son. I have to say, that Talib Kweli; the Reflection Eternal LP; when I think back on it, I actually stole that compact disck from my cousin and snuck and listened to it; because my mother didn't really like me listening to rap, even though there is a distinction between rap and hip-hop, and Talib Kweli was straight hip hop. I was deep into Wu-Tang. My pops is East Coast, and my mom is from the south, Louisiana. They both met in the capitol of California, then I, Jamel Jahi, started my cycle of life here on this planet. I'm amazed myself, but throughout my life I've just be in a warrior type of state. I've intensively studied my life, actions, choices, decision, etc, throughout all aspects thoroughly, almost to the point of obsession. I just want to rectify myself, and elevate to a supreme being. I love knowledge, it moves my very soul, and it tickles me so much to even get up out of my sleep at night to go and Google something that sparks my curiosity. Like I said, I'm inquisitive… I don't really want to get into all that just yet, maybe another book, the raw stuff. Now, my focus is information. Why information, because information is wealth, power, and love. This is the key to all, yet why do most choose to be ignorant, especially my brothers and sisters. I'm frustrated from the lack of unity, inspiration, motivation, idealism, compassion, mercy, and love… Where did it go, and who took if it was taken, or did we as a people just totally disregard it? If so, what the fuck that's not cool. I'm writing this to explain the words to my thoughts, this moment and in past times throughout my life and leading up to my overstanding of such things to be able to feel confident enough to stand up instead of sitting down like most have decided to. Like I said, I'm filled with the spirit of a warrior destined for greatness, but yet at a time I was reluctant to go to war. The war is now, the battlefield is in your mind; my weapons are these words on these pages. Digest my thoughts to grasp a clear overstanding of the perspective, yet listen wholeheartedly, without prejudice, or preconceived notions of any sort. Now, it is time for me to strike controversy; one part of me doesn't want to but I know we need it as a people because we're in trouble. I don't mean trouble like ooh, spooky scary ghost, I mean; your consciousness has been altered so you're not yourselves. We're the Gods and Goddesses of our own circumstances, and the world is waiting for us as a nation of people to take our rightful place on the planet. I'm sure, no at least I hope, some of you have at least pondered how we got in this situation. I know a lot of you are not going to like what I have to say, but I'm not putting my words to paper for you to like, but I love you though, not emotionally, yet a part is emotion because I'm a human being and I'm passionate about our culture, history, information, planet, universe, etc., but the other part is my duty as a civilized human being. So, I will challenge you on what you believe, because we have a lot of mythology that our people have been shaking their head to but don't question. I guess it's like a dance, once you get in the groove you don't want to stop; but sometimes the beat skips and you have to be intelligent and recognize what's taking place.

The Mind

Everything depends upon the mind. The world we experience is the result of our karma, or actions, and all the actions of the body and speech, originate in the mind. In order to change our world we have to start by changing our minds. Some people think that the mind is the brain or some other part or function of the body, but this is incorrect. The brain is a physical object that can be seen with the eyes and that can be photographed or operated on in surgery. The mind, on the other hand, is not a physical object. It cannot be seen with the eyes, nor can it be photographed or repaired by surgery. The brain, therefore, is not the mind but simply part of the body. There is nothing within the body that can be identified as being our mind because our body and mind are different entities. For example, sometimes when our body is relaxed and immobile, our mind can be very busy, darting from one object to another. This indicates that our body and mind are not the same entity. Our body can be compared to a guest house and our mind to a guest dwelling within it. When we die, our mind leaves our body and goes to the next life, just like a guest leaving a guest house and going somewhere else. If the mind is not the brain, nor any other part of the body, what is it? It is a formless continuum that functions to perceive and understand. Because the mind is formless, or non-physical, by nature, it is not obstructed by physical objects. It is very important to be able to distinguish between disturbed states of mind from peaceful states. A state of mind can disturb our inner peace, such as anger, jealousy, and desirous attachment, these are called 'delusions'; and these are the principal causes of all our suffering. We may think that our suffering is caused by other people, by poor material conditions, or by society, but in reality it all comes from our own deluded states of mind. The essence of cleansing the mind is to reduce and eventually eradicate altogether our delusions, and to replace them with permanent inner peace. This is the real meaning of our human life. The essential point of understanding the mind is that liberation from suffering cannot be found outside the mind. Permanent liberation can be found only by purifying the mind. Therefore, if we want to become free from problems and attain lasting peace and happiness we need to increase our knowledge and understanding of the mind. The fact that our inner experiences of pleasure and pain are in the nature of subjective mental and cognitive states is very obvious. But how those inner subjective events relate to external circumstances and the material world poses a critical problem. The question of whether there is an external physical reality independent of consciousness and mind has been extensively discussed by philosophical thinkers. Naturally, there are divergent views on this issue among the various schools of thought. Some assert that there is no external reality, not even external objects, and that the material world we perceive is in essence merely a projection of our minds. From many points of view, this conclusion is rather extreme. Philosophically, and for that matter conceptually, it seems more coherent to maintain a position that accepts the reality not only of the subjective world of the mind, but also of the external objects of the physical world. Now, if we examine the origins of our inner experiences and of external matter, we find that there is a fundamental uniformity in the nature of their existence in that both are governed by the principle of causality. Just as in the inner world of mental and cognitive events, every moment of experience comes from its preceding continuum and so on ad infinitum. Similarly, in the physical world every object and event must have a preceding continuum that serves as its cause, from which the present moment of external matter comes into existence. In most literature, you'll find that in terms of the origin of its continuum, the macroscopic world of our physical reality can be traced back finally to an original state in which all material particles are condensed into what are known as space particles. If all the physical matter of our macroscopic universe can be traced to such an original state, the question then arises as to how these particles later interact with each other and evolve into a macroscopic world that can have direct bearing on inner experiences of pleasure and pain. The invisible workings of actions, or karma, are intimately linked to the motivation in the human mind that gives rise to these actions. Therefore an understanding of the nature of mind and its role is crucial to an understanding of human experience and the relationship between mind and matter. We can see from our own experience that our state of mind plays a major role in our day-to-day experience and physical and mental well-being. If a person has a calm and stable mind, this influences his or her attitude and behavior in relation to others. In other words, if someone remains in a state of mind that is calm, tranquil and peaceful, external surroundings or conditions can cause them only a limited disturbance. But it is extremely difficult for someone whose mental state is restless to be calm or joyful even when they are surrounded by the best facilities and the best of friends. This indicates that our mental attitude is a critical factor in determining our experience of joy and happiness, and thus also our good health. Now, the mind can be defined as an entity that has the nature of mere experience, that is, clarity and knowing. It is the knowing nature, or agency, that is called mind, and this is non-material. But within the category of mind there are also gross levels, such as our sensory perceptions, which cannot function or even come into being without depending on physical organs like our senses. And within the category of the sixth consciousness, the mental consciousness, there are various divisions, or types of mental consciousness that are heavily dependent upon the physiological basis, our brain, (i.e. pineal gland, hypothalamus gland, pituitary gland) for their arising. Another distinctive feature of mind is that it has the capacity to observe itself. The issue of the mind's ability to observe and examine itself has long been an important philosophical question. In general, there are different ways in which the mind can observe itself. For instance, in the case of examining a past experience, such as things that happened yesterday you recall that experience and examine your memory of it, so the problem does not arise. But we also have experiences during which the observing mind becomes aware of itself while still engaged in its observed experience. Here, both observing mind and observed mental states are present at the same time, the phenomenon of the mind becoming self-aware, being subject and object simultaneously, through appealing to the factor of time lapse. It is important to understand that when I talk about mind, I'm talking about a highly intricate network of different mental events and states. Through the introspective properties of mind we can observe, for example, what specific thoughts are in our mind at a given moment, what objects our minds are holding, what kinds of intentions we have and so on. In a meditative state, for example, when you are cultivating a single-pointedness of mind, you constantly apply the introspective faculty to analyze whether or not your mental attention is single-pointedly focused on the object, whether there is any laxity involved, whether you are distracted and so forth. In this situation you are applying various mental factors and it is not as if a single mind were examining itself. Rather, you are applying various different types of mental factors to examine your mind. In our own day-to-day experiences we can observe that, especially on the gross level, our mind is interrelated with and dependent upon the physiological states off the body. Just as our state of mind, be it depressed or joyful, affects our physical health, so too does our physical state affect our mind. There are specific energy centers within the body that have some connection with what some neurobiologists call the second brain, the immune system. These energy centers play a crucial role in increasing or decreasing the various emotional states within our mind. It is because of the intimate relationship between mind and body and the existence of these special physiological centers within our body that physical yoga exercises and the application of special meditative techniques aimed at training the mind can have positive effects on health. It has been shown, for example, that by applying appropriate meditative techniques, we can control our respiration and increase or decrease our body temperature. Furthermore, understanding the subtle relationship between mind and body, we can practice various meditations while we are in dream states. The implication of the potential of such practices is that at a certain level it is possible to separate the gross levels of consciousness from gross physical states and arrive at a subtler level of mind and body. Our mind is divisible into two parts – the conscious and the subconscious. The conscious mind is the master and the subconscious mind is the servant. If the conscious mind holds on to a certain thought, the subconscious mind takes it as an instruction and proceeds to manifest into the being of our affairs. The conscious mind can search for facts on which to build its reasoning but the subconscious mind does its reasoning exclusively with the thoughts handed down by the conscious mind whether they are facts or not. So if you feed a constructive thought, you are giving a definite instruction to work positively. The function of the subconscious mind operates with and aside from that of the conscious mind. From this part of the brain, we are able to delve into realities beyond the physical aspects of existence to that of extra sensory perception, mind reading, being able to view future events before they happen and the very proficient individuals who are able to project thoughts into the minds of others. In most individuals, these abilities are chaotic and sporadic to say the least. In a few certain people, these abilities are more natural and easier to develop and control. The subconscious is the last physical function of the brain before access to the realization of the spiritual universe, which is the part of the mind, which holds the information of our very existence since the genesis of the Universe. Here we access knowledge from realization of who we are and the reasons for being. We also access the collective awareness from every living being on this planet and that of intelligence from beyond our galaxy and dimension. In other words, we become one with every soul, spirit and the creator. Each human possesses this knowledge as part of his or her relation to this ageless existence. It is the clutter of the physical world, which keeps us from realizing this.

Emotions

Many people can seem very sure of themselves while they are predominantly living in a state of unreality, and some people may doubt themselves for feeling confused about things, when in fact they are much closer to the truth than the falsely confident are. One cannot skip steps to arrive at clarity, and one cannot get there simply by acting sure of themselves. One way that I've conceptualized the process of personal evolution has been as a kind of three-part movement, that being from ignorance to confusion to clarity. Ignorance was my frame of mind in which others' personal belief systems were so firmly in place and embedded so far down in my subconscious mind that at a point I felt certain of my own validity, rarely questioning it. There are many such axioms or beliefs with variations and derivatives aplenty. Since these beliefs were not doubted for the most part, I sought out ways to better cope with what is accepted as the harsh realities of my life. Much energy was being invested in trying to find better and better ways of manipulating the self and other people to get what I wanted. To that end, I had developed what theorists have referred to as personas, or false selves, constructed to present to the world in order to attain sought after praise, recognition, love or substitutes for those things. Indeed, I was looking to polish up my mask so I might work better, and was surprised to find out that a key part of the real self work is, in fact, to expose and take off my mask. One of the first endeavors in my unfolding process was to begin uncovering the embedded beliefs behind my masks, and challenging their validity, thereby confronting the false clarity they offered as a substitute for knowledge and security. It is not easy. Indeed, if you believe something so thoroughly, you will invite, create or only be able to see that very thing in your life most of the time, so its reality will seem absolute. I invested in particular methods of trying to attain a modicum of happiness, I would not readily forgo. The strategy that I'd devoted much of my life to was actually faulty; a heartbreak. For the most part, in love relationships I've found myself feeling emotionally and intellectually deprived. In spite of tireless efforts to be agreeable, accommodating and self-sacrificing, I was just not getting enough intellectual stimulation, support, appreciation or affection from my partner. I felt so defeated and frustrated because while I believed that there really wasn't enough love in the world to go around, I was sure that the way to get what is available is by being agreeable, accommodating and self-sacrificing. What I was in denial of is the fact that my behavior was part of a mask, attempting to hide a very demanding and childish attitude towards my loved ones based on a buried belief in deprivation. My partner being bombarded with these masked demands would often withdraw and indeed be less inclined to give. So therefore, I had to manipulate even more, all the while building up a stockpile of resentment, on and on; in a self-fulfilling vicious cycle. The failure of the manipulations to get more of what is wanted is often what brought me to seek to find out what I'm doing wrong, why the mask I'm sure is based in reality is not having its designed effect. This was my life in a state of ignorance. Yes, you're right, sure you know how life works, but inexplicably unhappy, which, if the beliefs are seen as clearly right, can only lead people to the conclusion that they are failing, or manipulating inadequately. This is a place where the self is so rigidly defined according to firmly held beliefs that only lead to frustration and a sense of inadequacy? When blocked feelings are released through an integrated mind-body-spirit integration, the embedded beliefs start appearing in higher relief. My beliefs were previously being used to justify keeping emotions trapped in the body. As my emotional channels were opened, the old beliefs became subject to challenge and dismantling. I found myself at that point without the familiar, stereotypical ways of viewing the world, myself and others. Meaning, without the illusions created by projecting static images into the future, what can one count on for predictability? Again, the reason I was in this state of ignorance in the first place is because I realized that my life has never successfully followed my projections anyway, and the fulfillment promised by the illusions always seemed to remain unattainable or just out of reach. Now, I am starting to realize that I didn't know what I thought I knew. Now, I am in a state of confusion; it is here, at the I don't know who I am place that true wisdom began. What kept me going at this stage, fortunately, is that despite the confusion, I felt better, and often, in spite of an apparent lack of direction, my outer life was frequently improving. For one person, it may be physical health that improves, for another financial abundance arrives, or work-life becomes more creative. For still others; they break through a relationship barrier. Yet, for all, it is really the new inner feeling of self-possession and inner connectedness that provides motivation. At this stage of the process, I was able to uncover beliefs in deprivation and scarcity of love that's been embedded in my subconscious mind since early in childhood, based on a less-than-fully-gratifying relationship with a parent and family members. My feelings stored in my body since that time, the hurt and rage, have also now been energized through the ignorance and partially released at this stage as well. The origins of my masks have been uncovered and seen as primitive attempts to get more from a parent - with poor results, of course. It is now becoming understood that the world of love has been viewed through this tainted lens since my infancy. It is experienced as a revelation to consider that one could be fully gratified in an adult relationship, could give and receive all of the love that one is capable of without having to do anything to get it. It is also startling to realize that I have discounted or ignored the possibilities for greater love because to see that would have run counter to the absolutely certain beliefs that I was holding onto. This was the point at which the resistance to being wrong about one's strategy for living got confronted. It was painful. I was faced with the fact that all of the feelings of failure and frustration, and of course, self-hate, were not based on reality at all, but on an erroneous conclusion about life which originated in early childhood. Great courage was required to forge on here. When my deep primal feelings had been to a great extent released, and the core negative beliefs very much unveiled, I came to a new place of overall inner security and openness that provided both confidence to trust what I know in the moment, and simultaneously, flexibility to re-evaluate my knowledge and change when called to do so; this is clarity. When a person knows how they feel in the moment, and is aware of their immediate inner thought processes as well. Judgmental attitudes about emotions and the contents of one's mind are not held onto. At this stage I started to follow my instincts in major decisions without a lot of second-guessing or rigidly gripping to projections and anticipated outcomes. When one is in a state of clarity, the truth of matters is no longer mainly sought through deductive reasoning, but rather through inner resonance with the truth, and actions are decided upon by trusting gut feelings; my mother taught me that. Anyways, I now know that whatever transpires, I will be open and flexible enough to creatively move with the events. Mistakes and temporary obstacles are accepted as information, not measured against images of perfection or rigid beliefs about success or failure.

I can experience the joy of being wrong in this state, that is the freedom from needing to come up with the right strategies, free from worrying about blowing it when making decisions, etc. For example, people who once believed in deprivation now know through experience that life is abundant with opportunities to exchange love and pleasure with another and that the only efforts one must make to that end are to keep the emotional channels in one clear, or balanced. Good things seem to just arrive as a by-product of being more genuinely oneself. The person understands, too, that we all act like magnets for experiences in life, and that we will attract whatever we are charged up with. If it is joy and love that we are energized with, we will attract joy and love. If it is hostility, likewise that is what we will attract. So, unpleasurable events are dealt with by going within to examine one's inner state. I am also clear now that one's attitude towards oneself is one's attitude toward others.

Religion

To overstand the background of religions, one must appreciate the basis of all ancient religions, going back to the Phoenicians, the Babylonians, the Sumerians, and beyond the Indian backdrop of religion, the Oriental backdrop of religion, and the Semitic backdrop of religion. In doing so, we must study linguistics and etymology. Now, the Indians oldest dated language is Sanskrit, which gave birth to Hindu, Adamic (Cuneiform and Sanskrit), Ugaric (Hindi and Cuneiform). Ugaric, later gave birth to Akkadian, and Armenian also known as Nordic, another Aramaic dialect, gave birth to the Indo-European languages. Indo-European languages are the languages of Europe and southwest Asia and India. The Persian language, also called Farsi, is considered part of the Aryan or Indo-Aryan languages. However, Persian has been divided into three different languages for three different eras of time. Old Persian was written in Cuneiform; Middle Persian used Aramaic; and Modern Persian is what the Arabs call Arabic today. The Oriental's oldest dated language is Chinese which gave birth to Japanese and Korean and many sub-dialects mixing with Euro stock produced dialects as Altaic, Manchu and Sino-Tibetan. The Semitic's oldest dated language is Ugaric which gave birth to Ashuric and Akkadian, and many sub-dialects mixing in with Euro stock produced the dialects as Aramean, Chaldean, Hebrew, and Arabic. All these languages used the script known as Cuneiform, which is a written and spoken language of the Neteraat, which is the only language not mentioned in the bible by name. Cuneiform is the oldest written and spoken language on the planet earth which was used by the Pre-Dynastic Neolithic Egiptians. When it comes to religion, the game that was played was to incorporate an existing culture that can verify itself and give itself some validity to say through each one of the major prophets that they interacted with Egipt. Egipt still stands today, and it is a fact that it exist. The pyramids are there. These so-called religious scholars haven't been able to verify any of the stories in the bible. Go to Jerusalem and find Jericho; they can't find it. They haven't found the Garden of Eden yet. They haven't even found were Noah's ark landed. They haven't found the Ark of the Covenant. So these people who found all of these religions and ideologies aligned themselves with something everybody did know, which was Egipt; to give themselves validity. I'm saying that to say that the Jews; the Torah, aligned themselves with Egipt because Egipt was real. It makes them look real. If you strip away the connection to Egipt there is no other culture to verify your bible. Yashua was hung on a cross; the concept of crucifixion is found in Egipt. The ankh was the original cross long before the crucifix was introduced into Christianity; or the fish. The fish of Christianity goes back to Dagon; a Sumerian god that came out of the Ogdoad, which were the Egiptian deities of the sea. Once again, Egipt is there every step of the way. They fear you becoming conscious of that. They don't need you waking up to that reality. They don't have Solomon's temple. Saddam & Gomorrah; they never found it. They have no proof of any of this stuff. It's all symbolism. If you go down to South America you can see the Mayan pyramids; they have a calendar, names of gods who talk about creation, etc. How hard is it to go to Egipt and find the pyramids, the obelisk; they're there, and they predate Yashua. The sphinx is older than Yashua but its there. Can you give me any proof that Yashua existed two thousand years ago other than faith and belief? Remember that hearing and reading is believing, and seeing is knowing. That is a fact, right knowledge.

The problem is when you're asking me to have faith and belief when I can find facts. How could I accept these images as god when I can find facts of my ancestors as gods, people as gods and their descendents as gods. I can go to Egipt and look on the wall and see my face, your face, etc. They made sure they made replicas of everybody; that's what there eternal life is about. You can't even tell me for sure what Yashua looks like. You don't have one face of a man two thousand years ago and I can show you about a thousand faces of Ramses four thousand years ago. Well actually, they say Ramses existed twenty thousand years ago; ok it's still before Yashua, so his face is there. Where is Yashua' face? People back then were making faces (gene splicing). If God was here, don't you think they'd make his face? So now, we got the Shroud of Turin. The story is that Yashua was put inside of a tomb, and a stone was rolled in front of it. Yashua then resurrected through the cloth that they wrapped him in and left the impression of his face on the cloth. But why didn't he leave his impression on the stone when he penetrated the wall. So Yashua resurrected in the spirit but they found the stone moved, why? The spirit can walk right through a stone; if the stone was moved, Yashua walked out physically or some people went in and carried him out physically. Why would he leave his impression on a cloth? There explanation is because he wanted people to know what he looked like. Yashua could step back through time today and say this is what I look like and it would be convincing, a lot more persuasive than a cloth. Did he do it? Yes. Does God do that, yes. Angels step in the world get into mans affairs and step out. God did it. God came down to earth and confounded their tongues in Babel, and then went back up to heaven. How do we know he went back up to heaven; because Yashua said so in Matthew? He said " I want you to pray like this, our father who art in heaven." He told people where God was; which established where God wasn't. He told us where God was, and in doing that he told us where God was not. Now there is not a place where my God isn't. You must have a different god. Your god could be in a place; my God is the place. Your god could be in heaven; my God is heaven. My God doesn't go, he is. My God doesn't come, he is. Do you see the difference? The definitions as taught through the Bible; through Christianity, Judaism, and Islam is to reduce the power of God cause the next step to God is angel and the next step to angel is man. If you reduce the power of God, you lose all contact with angels. So if you bring God down and angels down; in doing so, man becomes this weak unimportant subservient creature that's constantly telling God help me, help me. God doesn't help anybody that doesn't help themselves. If you want a helping hand you'll find it at the end of your arm; if you don't get it done, it won't get done. You have the power to make things happen and they cut that off. You have the power to think things into being. They made you doubt your own powers. You are like Yashua; you are a Yashua. Yashua said so. Now first, you have to prove that Yashua exists, because the burden of proof is on you because that's your religion. I can prove Osiris existed; if not as a man, then as a rock. You can see his face in Egipt on a rock, and that rock exist; not only in concept, but in reality. These religions and dogmas have you working with a concept god, not a reality. When I ask you to present to me or bring me to the world of reality, you get mad at me because you can't do it. You start calling me names. Now let me tell you this; the most books written in this country in English are cookbooks and books on Egipt (the most various by different authors). The Freemasons, Shiners', Knights of Templar's, Illuminati, Klu Klux Klan, etc; all of these are institutions trying to get to the root of Egipt. Now are you ready to learn the truth, or do you want to live in the lie; the game of portraying who wants to be an Egiptian. They want to be you. Meanwhile, you are trying to be them. Find the truth of yourself, and you will find God; in yourself. The power of God is you; all in, around and through you. They don't want this information out, cause there is no place in the darkness for light without chaos. The moment they get involved, trouble starts. When they came to Egipt, trouble started, when they came to America, Africa; trouble started. Look at Australia, Trinidad, China (Marco Polo). When Marc Anthony and Julius Caesar went to Egipt, trouble started. Wherever they go; they are the brothers and sisters and son and daughters of chaos. It's time for us to take back what was ours. America is our land, not just Africa. This is our land, we were living here on these shores and they came seeking gold and massacred us for it; took over, and now got us going to court saying 'In God We Trust.' Which god are you talking about? In this country there's Muslims, Christians, and Jews; all having three different gods. There's no God in the Christian Bible; that's a Greek translation. 'I swear to tell the truth in the name of God' what is the name of God? Yashua, Eloheem, Adonai, use the name of God. They got us using God. Gomer (wisdom), Oz (strength), and Dober (beauty); when that doesn't define anything but man. They say man is made from the ground. You are in tune with the sun, the water, and you're the color of the ground. We had gold, we had diamonds, we had oil, petroleum, all the natural resources that they call wealth; growing from where we came from. There plan is to sedate you and I; if not with drugs, they'll sedate you with religion. Both of these things turn us against each other. People argue all day about Yashua and Allah, while the governors of these religions and dogmas sit back and laugh at you and I. You are arguing about our ancestors as Gods. They gave us garbage to eat and call it food, or nutrition. They have the upper hand, the psychological entrapment of your mind called fear of the unknown. Like the fear or paranoia you feel about what they are going to do to you if you emancipate yourself from faith and belief; fiction. They tell you you're going to hell, etc. That's because you are not in control; if you get back in control of things, you make the decisions, or have the power to make things happen. Thought or your intellect is the switch that controls your ability to tap into that universal power that can transform visualization into a reality. You have that power. You have lost contact with who you are. This doesn't mean I don't believe in Yashua; but he's a Moor just like me.

That's even what the Bible says. I can show you in the Bible in Revelations were Yashua is a Moor. For the record, Moses was an Egiptian. I want you to show me somewhere in there were he's not. If you can't; tell me why it's wrong for me to worship God like me? If you go to China, people in China worship gods that are Chinese. If you go to India, they worship gods that are Indian. If you go amongst the Arabs, Muhammad is an Arab. Now when it comes to me and you, they want us to worship gods that look like them. We are the only people that surrender to that. Let me ask you a question. Is the spirit of God in you or not? Is the light of God the light in man? So, are you a God? Some say God is in me, or in Christ, Muhammad, Buddha, Osiris, Tehuti, etc. If God is in all things and out of all things simultaneously; then what makes it wrong when you and I say that we are Gods? You think this is a coincidence? There is incidence, coincidence, and reality. An incident occurs once, coincidence is twice, and anything that happens three times is a reality. These people that teach you and I to watch out for the devil are teaching the message of the devil. Every time they tell you to be good, they are making you remember that there is a bad. They are teaching the evil as they are teaching good. That is the devil's mission, his philosophy. Give it all back. Get involved with yourself, and your family. Make a spiritual and blood link to your ancestors. No one should be able to tell you not to worship your ancestors and Yashua is worshiping his ancestors. Look in the book of Exodus. Now when you or I say Osiris, Shu, Tefnut, something is wrong; I'm called a pagan, idol worshiper and polytheist. They call you everything except what you are. Your Torah, Bible, and Quran stories all come from Egiptian stories changed to psychological dull your mind. They knew that if they told you stories about yourself, but with new names and faces, they could capture your soul and call it faith and belief. That is not to condemn all people who call themselves Christians, Jews, Muslims, etc. There are many who express loving spirituality through their beliefs. I am referring to the institutions of monotheism and its arrogant indoctrination and imposition of its desperately limited vision of life which has created a prison for the mind for literally billions of people for years. Now, to further clarify, The Zodiac sign Capricorn, which on December 22 -25, during the Winter Solstice (the day the Sun stood still) for three days, the Sun reaches its lowest point in the skies and on December 25 began its apex climb to be born nine months later in the month of September under the Zodiacal sign of Virgo, the Virgin and planetary influence of Venus, symbolizing love. This is the same exact story of Mary, the Virgin Mother given birth to the Son (Sun). During December 22-25th the Sun (Son) is in the Tomb or the Grave, like Yashua is the Son (Sun) of God and laid in the tomb for three days and three nights and on the third day he resurrected from the tomb, having the keys over death and hell. Therefore, the Zodiac sign Capricorn represents descending energy or light (Spirit) condensation into Matter ($E = Mc2$). This also symbolizes the so-called fall of man and why Yashua was also called the Son (Sun) of man (Mind). Lucifer Is the encasement of star fire, called the Kundalini or serpentine fire in the human body. Lucifer makes his appearance in the fourteenth chapter of the Old Testament book of Isaiah, at the twelfth verse, and nowhere else:

> *"How art thou fallen from heaven, O Lucifer, son of the morning! How art thou cut down to the ground, which didst weaken the nations!"*

The first problem is that Lucifer is a Latin name, so how did it find its way into a Hebrew manuscript which was written before there was a Roman language? Yes, Lucifer is a Latin word, however, it is derived from the Hebrew word Uriel, which means light of my God and

Uriel was one of the seven Arch-Angels or Eloheem (group of gods or angels). Uriel's ancient Egiptian name was Heru, the God of light. Heru's other name was Ma Khu Ra or its Hebrew transliteration Mi-Cha-El or Michael. Therefore, Uri-El and Mi-Cha-EL are two different extremes of the same polarity, which actually, symbolizes the seven degrees of consciousness or the seven states of consciousness; spirit (light energy) to matter (crystallized light). The _"BOOK OF ENOCH"_ (or Adafa) names seven archangels: Uriel, who rules the world and Tartarus (Hell); Inter-Personal Consciousness. Raguel, who takes vengeance on the world of the luminaries; Intra-Personal Consciousness. Saraquael, who is set over the spirits; Life Consciousness. Raphael, who rules the spirits of men; Sub-Consciousness. Ramiel, whom God set over those who rise; Super Consciousness. Gabriel, ruler of paradise, the serpents and the cherubim; Magnetic Consciousness, and Michael, who is set over humanity and man-kind and chaos; Infinite Consciousness. In the original Hebrew text, the fourteenth chapter of Isaiah is not about a fallen angel, but about a fallen Babylonian king, who during his lifetime persecuted the children of Israel. It contains no mention of Satan, either by name or reference. The Hebrew scholar could only speculate that some early Christian scribes, writing in the Latin tongue used by the Church decided for themselves that they wanted the story to be about a fallen angel; a creature not even mentioned in the original Hebrew text, and to whom they gave the name Lucifer. Why Lucifer? In Roman astronomy, Lucifer was the name given to the morning star (the star we now know by another Roman name, Venus). The morning star appears in the heavens just before dawn, heralding the rising sun. The name derives from the Latin term lucem ferre, bringer, or bearer, of light. In the Hebrew text the expression used to describe the Babylonian king before his death is Helal, son of Shahar, which can best be translated as "Day star, son of the Dawn." The name evokes the golden glitter of a proud king's dress and court (much as his personal splendor earned for King Louis XIV of France the appellation, "The Sun King"). The scholars authorized by King James I to translate the Bible into current English did not use the original Hebrew texts, but used versions translated largely by St. Jerome in the fourth century. Jerome had mistranslated the Hebraic metaphor, "Day star, son of the Dawn," as Lucifer, and over the centuries a metamorphosis took place. Lucifer the morning star became a disobedient angel, cast out of heaven to rule eternally in hell. Theologians, writers, and poets interwove the myth with the doctrine of the fall, and in Christian tradition Lucifer is now the same as Satan, the Devil, and ironically the Prince of Darkness. So Lucifer is nothing more than an ancient Latin name for the morning star, the bringer of light. That can be confusing for Christians who identify Christ himself as the morning star, a term used as a central theme in many Christian sermons. Yashua refers to himself as the morning star in Revelations.

> *"I Yashua have sent mine angel to testify unto you these things in the churches. I am the root and the offspring of David, and the bright and morning star."*
>
> *(Revelations 22:16).*

This simply shows that Lucifer, the Light Bringer or the Light Bearer is the same as Yashua, the Light of the World or the Sun (Son) of God. Ponder that!

> *"And we have the word of the prophets made more certain, and you will do well to pay attention to it, as to a light shining in a dark place, until the day dawns and the morning star rises in your hearts."*
> *(2 Peter 1:19).*

"Just as I have received authority from my Father, I will also give him the morning star."

(Rev. 2:27-29).

May its morning stars become dark; may it wait for daylight in vain and not see the first rays of dawn, for it did not shut the doors of the womb on me to hide trouble from my eyes."

(Job 3:1-10)

"Where were you when I laid the earth's foundation?... On what were its footings set, or who aid its cornerstone - while the morning stars (meaning more than one morning star) sang together and all the angels shouted for joy?"

(Job 38:4-7).

This simply demonstrates that the characters Lucifer (Set-Heru, Set-An or Satan, UriEL) and Yashua (Isau- Heru, MichaEL, Ma Khu Sut-Tekh or Mel-Chi-Sedek) symbolizes one of the Seven Universal Laws of Tehuti; Polarity.

The Seven Principles of Tehuti

1. **Polarity**
   "Everything is dual; everything has poles; everything has its pair of opposites; opposites are identical in nature, but different in degree; extremes meet; all truths are but half truths."

2. **Mentalist**
   "Everything is mental; the Universe is a mental creation of the All."

3. **Correspondence**
   "As above so below; as below so above; as within so without; as without so within."

4. **Vibration**
   "Nothing rests; everything moves; everything vibrates.

5. **Rhythm**
   "Everything flows out and in; everything has its tides; all things rise and fall; the Pendulum-swing manifests in everything; the measure of the swing to the right is the measure of the swing to the left; rhythm compensates."

6. **Cause and Effect**
   "Every Cause has its Effect; every Effect has its Cause; everything happens according to the law; chance is but the name for a law not recognized; there are many planes of causation, but nothing escapes the law."

7. **Gender**
   "Gender is in everything; everything has its Masculine and Feminine Principles; Gender manifests on all planes."

Henry Neufeld (a Christian scholar comments on this Biblical issue) went on to say,

> *"And so there are those who do not read beyond the King James Version of the bible, who say 'Lucifer is Satan: so say the Word of God'..." "This passage is often related to Satan, and a similar thought is expressed in Luke 10:18 by Yashua, which was not its first meaning. It's primary meaning is given in Isaiah 14:4 which says that when Israel is restored they will "take up this taunt against the king of Babylon . . ." Verse 12 is a part of this taunt song. This passage refers first to the fall of that earthly king... How does the confusion in translating this verse arise? The Hebrew of this passage reads: "heleyl, ben shachar" which can be literally translated "shining one, son of dawn." This phrase means, again literally, the planet Venus when it appears as a morning star. In the Septuagint, a 3rd century BC translation of the Hebrew scriptures into Greek, it is translated as "heosphoros" which also means Venus as a morning star. How did the translation "lucifer" arise? This word comes from Jerome's Latin Vulgate. Was Jerome in error? Not at all. In Latin at the time, "lucifer" actually meant Venus as a morning star. Isaiah is using this metaphor for a bright light, though not the greatest light to illustrate the apparent power of the Babylonian king which then faded."*

Lucifer represents the Kundalini and its color is red, located at the root base chakra; red is the color of the devil or double of the Spiritual Body, the Physical Body, which symbolizes Satan. Lucifer wasn't equated with Satan until after Jerome. Jerome wasn't in error.

Later Christians and Mormons were in equating Lucifer with Satan. So why is this a problem for Christians? Christians now generally believe that Satan or the Devil or Lucifer who they equate with Satan is a being who has always existed or who was created at or near the beginning. Therefore, they also think that the prophets of the Old Testament believed in this creature. The Isaiah scripture is used as proof and has been used as such for hundreds of years now. So why is Lucifer a far bigger problem to Mormons? Mormons claim that an ancient record was written beginning in about 600 BC, and the author in 600 BC supposedly copied Isaiah in Isaiah's original words. When Joseph Smith pretended to translate the supposed ancient record, he included the Lucifer verse in the Book of Mormon. Obviously he wasn't copying what Isaiah actually wrote. He was copying, the King James Version of the _"HOLY BIBLE."_
The author of _"THE POLYTHEISM OF THE BIBLE AND THE MYSTERY OF LUCIFER,"_ F.T. DeAngelis, comments on this page as follows…

> *"It seems minor, but - the actual term used in the Greek Septuagint version of Isaiah 14:12 (given that there is no one way of accurately transliterating) is Eo(u)s phoros, morning star/DAWN god of light. The actual name, "Lucifer," goes*

> *back to the Greeks, before the Romans. Socrates and Plato talk about this "god of light"; surprisingly, not in the context of Eos (god of Dawn), but -- as a morning star -- juxtaposed with the sun (Helios) and Hermes. This information can be found in Plato's <u>"TIMAEUS"</u> and in Edith Hamilton's <u>"MYTHOLOGY"</u>.*

On a lighter note, Arthur Clarke, in his fictional book <u>"2061"</u> correctly uses the word Lucifer. He uses it as a name for a new sun in the solar system which is correct since the new sun is a second 'morning star' of 'original' 'light-bearing' substance--not some evil being of religious mythology. David Grinspoon comments on the historical aspects of the word as follows:

> *"The origin of the Judeo-Christian Devil as an angel fallen from heaven into the depths of hell is mirrored in the descent of Venus from shining morning star to the darkness below. This underworld demon (daemon, demi-urge), still feared today by people in many parts of the world, is also called Lucifer, which was originally a Latin name for Venus as a morning star. <u>("VENUS REVEALED" p. 17)</u>.*

Actually, Grinspoon should just refer to the "Christian Devil" since the Jews never believed in such a creature and still don't to this day.

> *"When the Mason learns that the Key to the warrior on the block is the proper application of the dynamo of living power, he has learned the Mystery of his Craft. The seething energies of LUCIFER are in*

> his hands and before he may step onward and upward, he must prove his ability to properly apply [this] energy."
>
> - *Lost Keys of Freemasonry*, Manly P. Hall

> 'Get down (from the Garden or Higher Mind State), one of you an enemy to the other (i.e. Adam [Left Hemisphere], Eve [Right Hemisphere], and Satan [Kundalini Shakti]). On Earth (Body) will be a dwelling-place for you and an enjoyment -- for a short time'. He (ALLAH, God) said:' Therein you shall live, and therein you shall die, and from it you shall be brought out [i.e. resurrected].'
> - (Quran 7:24-25).

Adam and Eve were ordered to leave the Garden which they were in and descend to Earth where they and their children will live and die, and where Satan will also be. The *"BOOK OF ENOCH"* says,

> "there were angels who consented to fall from heaven that they might have intercourse with daughters of earth. For in those days the sons of men having multiplied, there were born to them daughters of great beauty. And when the angels, or sons of heaven, beheld them, they were filled with desire; wherefore they said to one another: 'Come let us choose wives from among the race of man, and let us beget children.' Their leader Samyasa, answered thereupon and said: 'Perchance you will be wanting in the courage to fulfill this resolution, and then I alone shall be answerable for your fall.' But they swore that they would in no wise repent and that they would achieve their whole design. Now there were (200) two hundred who descended Mount Harmon, this

> *metaphysically means, the 200 Fallen Angels actually symbolizes the 200 solidified bones of an adult human skeletal system. In astrology, the bone or skeletal system, DNA, and the blood are helped formed by the planetary energies of Saturn and by the cosmic energies of the constellation Sirius. In the <u>"EL SUHUFAN,"</u> part of the <u>"HOLY TABLETS"</u> by Dr. Malachi Z. York EL, it states that the after the flood of Noah, they left the planet Earth and went to reside on the planet Saturn), and it was from this time that the mountain received its designation, which signifies Mount of the Oath. Hereinafter, follow the names of those angelic leaders who descended with this object: Samyasa, chief among all, Urakabarameel, Azibeel, Tamiel, Ramuel, Danel, Azkeel, Sarakuyal, Asael, Armers, Batraal, Anane, Sameveel, Ertrael, Turel, Jomiael, Arizial. They took wives with whom they had intercourse, to whom they also taught Magic, the art of enchantment and the diverse properties of roots and trees. Amazarac gave instruction in all secrets of sorcerers; Barkaial was the master of those who study the stars; Akibeel manifested signs; and Azaradel taught the motions of the moon."*

The Planet Saturn, the sixth planet from the Sun, is known in the occult world as the "Black Planet," therefore, it metaphysically symbolizes Melanin. It is also called the "Eye of Buddha (Putah, Ptah)," the "Eye of Brahma" or the "Eye of Christ".

> *The light of the body is the eye: if therefore thine eye be single, thy whole body shall be full of light.*
>
> *- <u>Matthew 6:22</u>*

> *But if thine eye be evil, thy whole body shall be full of darkness. Therefore the light that is in thee be darkness, how great [is] that darkness!*
>
> - <u>Matthew 6:23</u>

When it comes to beliefs, etc., always remember that the ghost man can only reflect the light of the sun. He has no light or sun of his own, thus he is a fake man or human being with fake intelligence. They take the real and make the unreal, like plastic and seedless grapes. They take the facts like the existence of dinosaurs and create the fake dragons, Godzilla, Barnie; and the true fact of history is that ancient Egiptian became the myth and fiction in Judaism, Christism, and Muhammadism. These religions became real without any truth or facts that are unseen, unproven gods or god who himself had to have a physical son, Yashua, to make the unreal real. The ghost deity with no real purposes becomes a judgment god who kills, burns, destroys and turns people into stone, all without any true existence or reason. He is a ghost. The unreal becomes real faith and belief, so the fake man, which is the original man's kind or mankind is also fake versus the hu(e)man or human being. So he and she live in a fake world of movies, plays, fake food, and stimulants. It's all fake, the unreal; an illusion. In Christian terms, the soul returns with its harvest of experience "bearing its sheaves with it" in the world of the dead, to sit on the chair with god, a ghost. Traditional eastern religions such as Buddhism and Hinduism are principally concerned with what one may term the withdrawal aspect of life through detachment of the spiritual principle from its involvement in the form or material aspect; self-humiliation, self abuse, starvation, disfigurement, and Islam is a mixture of all these fanatic beliefs. It has them all. These religions are mainly laying emphasis on the joy and bliss after death, an unconfirmed state of existence; a blind date, which is experienced when the soul is able to withdraw from its attachment to the personal self, and achieve union with the higher spiritual principle without the body; a heaven if you do what they say to do with your whole life, and a hell if you don't obey these books written by men; Torah, Bible, Quran, etc.. This cuts you off from your own divine expression; you being the God of your life and taking the responsibility for all that surrounds you. The word "personal" derives from the Latin persona, meaning a mask. Its use indicates the transitory or unreal nature of the lower self as being something which we put on at birth through which to express ourselves, but it is not our real self. Self surrender is an essential of all mysticism and the playground of insanity and fanaticism and all religions have expressed this concept. Other philosophies, particularly western ones, are more concerned with the aspect of bringing down the spiritual principle and expressing it in form through creative activity. None of these religions is wholly right, or rather, each expresses a partial truth and so each is right only within its own limitation and are necessary parts of the whole process of life, living, and learning. Life is therefore both descent from ones' source and return to it, the All. In reality, one should balance the two processes. These two aspects may be termed magical and mystical, the path of descent being magic and the path of return mysticism. Now, there are other views express that the Supreme Neter (God) was called RA, the Father of the Neteru-Gods, and the Ruler of the Universe, who constructed himself out of Chaos (Nun). Chaos was, therefore, prior to RA, as infinite potentiality, and RA was a self-restricted aspect of it, the intelligence that produced Order (Ma'at). Contrary to popular opinion, RA is not the Sun Neter (God), Atum. RA is the living force that dwells around all things and lives within all beings. He is what the Chinese call in their esoteric system Chi or Qi, (Ki) energy, the life force, and in other esoteric or Vedic teachings, he is (Shiva [Shakta], Shu, YaH-Shu-WaH or Sekhem). The feminine aspect, is known as Kundalini (Sekh-Ma'at). In essence, RA is androgynous energy. Holy is the English transliteration of the Latin word 'Halo', which means 'Sun' and 'Halo' is derived from the Greek word 'Helios' which, means 'Sun'. The ancient Egiptian word for 'Sun' was 'On' or 'Ra'. The word 'Bible' is derived from the Greek word 'Biblios' meaning "Little Book" or 'Pages' which is derived from the ancient Sumerian word "Bab EL," meaning "Door to GOD (Sun) " or "Gate To GOD (Sun)" and even further back to the Egiptian word 'Papyrus'. Thus, the <u>"HOLY BIBLE"</u> is the "On Papyrus" or the "Ra Papyrus," which is the <u>"COMING FORTH BY DAY[LIGHT] AND NIGHT [SHADOW]"</u> misnomer the <u>"BOOK OF THE DEAD"</u>.

> *"So God [the Black Dot, the Infinite Point] created man [Mind] in his own image [Universe], in the image of God created he him; male and female [androgynous] created he them").*
> - *King James Version, Genesis(1:27)*

The legends of RA are numerous as to his journey across the regions (constellations). His royal barque (boat) is likened to that of the journey of the Sun; he is raw solar energy. According to the Latin etymologists, we can overstand what RA really is; he is the RA in aura, the force in radiation, radiance, ray(s), the chakras; that bright solar force that emits from all living beings. RA gives us spiritual strength. The KA is usually translated as spirit. The KA or somatic ego serves to bridge the gap between the physical and the psychic. It can be considered both as the etheric principle - a sort of subtle or higher or more psychic grade of the physical. As a matter of fact, the etheric body is situated approximately two inches above the physical body, and is often times a bluish-white color. Khnum, the ram-headed Neter (God) crafted the KA (Spirit, or Eve) & BA (Soul, or Adam) on his potter's wheel before a person's birth. It was thought that when someone died they met their KA. The KA is the intellectual and spiritual power. Each person was born with his or her KA, which was a constant companion through life and lived on after death, returning to its etheric origin. Now that I have hopefully opened your eyes, some of your old beliefs, your old ceremonies, may seem to you absurd; perhaps, indeed they really are so. Yet though you can no longer take a part in them, respect them for the sake of those good souls to whom they are still important. They have their use; they are like those double lines which guided you as a child to write straight and evenly until you learned to write far better and more freely without them; but now that time is past.

History

Let's speak about history, meaning "inquiry or knowledge acquired by investigation" which is the study of the human past. It is a field of research which uses a narrative to examine and analyze the sequence of events, and it sometimes attempts to objectively investigate the patterns of cause and effect that determine events; it is a way of providing perspective on the problems of the present. The stories common to a particular culture, are usually classified as cultural heritage rather than the disinterested investigation needed by the discipline of history. Also keep in mind that events of the past prior to written record are considered prehistory. Knowing that, I ask you whose story are we talking about, or better yet, whose story are you living by? How do you know you are a human being? How do you know it's nighttime, etc. What I'm saying is who taught you what you know. Have you ever thought that maybe what you think you know needs to be thoroughly investigated, or let's say understood at the very least. So who are you? When I say that, the first thing that comes to my mind is Sho Nuff from "The Last Dragon"; he said

"Who Am I?" That makes me laugh every time. My mother use to ask me who I was when I was young, and I would say, "I'm Jamel." Thinking back on it, it's funny because I know now what she meant. Anyways, before going any further let me just say that there is no such thing as ancient black people. Black is an adverb not a noun, however, a description, it is not an identity. We are very comfortable with using it to communicate with each other in a manner in which we can understand or accept. But yet, this goes against logic, certainly for any leader, guide, scholar or teacher, because their duty and mission is to tell the people the truth, and guide them toward knowledge, and not just give them what feels good or comfortable to satisfy their emotions. This is not about emotions, it is about the facts. The people are able to grasp what is being said to them, especially after they learn words, and how to read. A child can understand that he/she is not black. In their innocence they would say: *"Mommy if we were black, wouldn't we look like my black shoe"*. Therefore, to say that the people won't understand is an insult to the people, or an excuse. A thinking mind of a child can see that confusion exist. At the very least, the variety of pseudo identities, supposedly for the collective, causes dis-unity for children as they grow up. This is partly why there is so much discord amongst each other regarding who we are. Without serious studies, and without having the keys, one will not be able to know, nor be able to guide anyone else. Most of us certainly feel the pain of the conquest intent, and most of us say we wish to do something about it, to be in the struggle for change. Our own actions, or lack of action, keeps us from being responsible to ourselves, to our children and to others. Ignorance of these facts does not make them an untruth, nor will the truth pass away. Seek ye' the truth, and all else will come unto you. When I say history, I mean Ourstory! Before the concept of slavery, the historical falsifications, and beliefs, etc. EL-Mu'uria or Mu is said to be the original parent of the earliest world cultures and this partly explains the amazing similarities between ancient world cultures. The legendary super civilization of EL Mu'uria (Lemuria, Mu) flourished over ninety thousand years ago before sinking beneath the Pacific Ocean, over twelve thousand and five hundred years ago during the melting of the last ice age. The Egiptian historian, Manetho gave a history of Egipt's dynasties that goes back at least 24,927 (approximately 25,000) years to the time of EL Mu'uria, or LeMuria, (Mu and Aztlan), or Atlantis. In Plato's _"REPUBLIC,"_ an Egiptian priest refers to the destruction of EL-Mu'uria, or LeMuria, (Mu & Aztlan), or Atlantis when the states, *"There have been many destructions of humanity... you (Greeks) remember a single deluge only."* There is abundant worldwide archaeological, mythological & geological evidence which proves that an advanced, ancient super civilization once thrived in the Central Pacific. Ancient underwater cities have already been discovered in the Pacific Ocean. The ancient Egiptians spoke of the *"Twa* [Ptah-ites,Tengu, Anu, Ainu and 'Pygmy'] *people"* -small brown men who were early inhabitants on Earth. The Ainus, Japan's oldest known inhabitants, have traditions which tell of a race of dark dwarfs which inhabited Japan before they did. Historian Albert Churchward saw the Ainus as originating in Egipt. There is archeological support for this. In addition, J.A. Rogers writes in "<u>SEX AND RACE Vol. I</u>," there are a number of Japanese who, but for color and hair, bear a striking resemblance to the South African Bushman..." The Bushmen also known as Khwe [Khoe], Basarwa, or San peoples of South Africa and neighboring Botswana and Namibia are part of the Khoisan group and are related to the Khoikhoi. They have lived in southern Africa for some twenty thousand years. Along with the Pygmy of Central Africa, the

Bushmen have been considered a possible root or source for the female DNA lineage—the so-called Mitochondrial Eve. An ancient tradition points to the conquest of Japan from the southeast by a race of Melanite warriors. Presenting a fascinating account of EL Mu'uria (Lemuria) and Aztlan (Atlantis), the inspired *"RIGHT USE OF WILL"* by Ceanne DeRohan, reports *"the Lemurians were small and brown."* In *"PYGMY"* by Kitabu Jean-Pierre Hallet, he documents the amazing Twa (Pygmies) of Zaire, as *"the world's most genetically pure ethnic group... surviving since the dawn of mankind in real harmony with God, nature and each other."* Twa (Pygmy) bones are found all over Earth. In Albert Churwards book *"SIGNS AND SYMBOLS OF THE PRIMORDIAL MAN"* it talks about the many ancient exoduses of Melanites out of Africa to other parts of the world. Some Africans must have returned to Africa since many African tribes assert that their ancestors came from the East [EL-Mu'uria (LeMuria) Mu]. Their descendants are said to be primarily the people of the South Seas and Oceania. In Hawaii, there's one island where only pure-blooded Hawaiians live. They have full Melanite features, dark skin & wooly hair! In *"100 AMAZING FACTS ABOUT THE NEGRO"* by J. A. Rogers, he states....

"...the people of ...Negro descent living in Asia and Oceania probably exceed in the number the present Negro population of Africa. The purest Negro types are in Southern Asia."

For example, the China's pyramids (Miru) are located near Siang Fu city in the Shensi province; the Chinese do not know how they got there and they were often mistaken for hills due to the erosion. According to *"THE GROWTH OF CIVILIZATION"* by J. Perry, he states that the Africans from the Nile Valley were the builders. The pyramids (Miru) in Japan are made of stones not indigenous to Japan. It is said they were built during the time of EL Mu'uria (LeMuria). As a matter of fact, the Japanese ancestry lies within African genes. An ancient tradition points to the conquest of Japan from the southeast by a race of African warriors. The Ganges, the sacred river of India, the Latin word "India" is from the Greek word Indus (or Indos) which means 'Black', is named after the Ethiopian (Ether Utopia[n] or Cushite [Cush means 'Black']) Ngu (King) Ganges that conquered Asia. Krishna, which means "The Black One"; it is derived from the ancient Egiptian word 'Karast' or "Ka-Ra-Sut" meaning "The Spirit of the Black Sun" which later was transliterated into Greek as 'Khristos' (Kristos) meaning "The Black

Anointed One" and into English as the biblical 'Christ' meaning "The Anointed". Buddha means "The Awakener" or "The Enlightener." It is derived from the ancient Egiptian name 'Ptah' [Ptah-Re] or 'Putah,' meaning "The Opener." The English transliteration of the word is Fath-er and other great sages arose from their successive civilizations, including the great King Asoka. The first people of India were Cushites (Ethiopians) from Africa. They even named the area the Indus-Cush Valley. These Africans founded the great Indus-Cush Valley Civilization around 3000 B.C.E. They brought with them many spiritual sciences which originated in Africa, such as those of Yoga, Kundalini, Reflexology and other so-called Eastern philosophies. Master Kilindyii's essay in *"AFRICAN PRESENCE IN EARLY ASIA"* by Van Sertima/Rashidi reveals the African roots of the martial arts (Mon-tu):

> *"Africans discovered very early that the movements of animals could be sued effectively to develop their fighting skills. Also, that 'animal principles' could be isolated within the consciousness and manifested into an unconquerable fighting force. The oldest records of kicking, throwing, wrestling, and punching techniques were found in Egypt (KaMA'at). These warrior scientists laid the foundation for all martial arts systems (Note: Originally, called Mon-tu), including Kung Fu, Judo and Karate".*

Karate supposedly, means "Empty Hands", however, it is really made up of ancient Egiptian words, such as, 'Ka' which means 'Spirit'; 'Ra' means 'Sun' and 'Te' means 'Life' or 'Living'. Thus, "KaRaTe" means "The Spiritual Sun of the Living." A reference to Chi, Ki or prana energy, called the Universal Life Force. Asia's African roots are well summarized in *"AFRICAN PRESENCE IN EARLY ASIA"* by Ivan Van Sertima/Runoko Rashidi, and *"AFRICAN PRESENCE IN EARLY CHINA"* by James Brunson. The original oriental people were African (Melanites or Blacks), and many of them still are Black - southern China and Asia. The earliest occupants of Asia were "small Blacks" (Melanites; 'Pygmies') who came to the region as early as 50,000 years ago. In *"THE CHILDREN OF THE SUN"* by George Parker, he writes

> "... it appears that the entire continent of Asia was originally the home of many black races and that these races were the pioneers in establishing the wonderful civilizations that have flourished throughout this vast continent."

Reports of major kingdoms ruled by Africans are frequent in Chinese documents. The first kingdom of Southeast Asia is called Fou Nan, famous for building masterful canal systems. Chinese historians described the Fou Nanese men as *"small and black."*

In "*100 AMAZING FACTS ABOUT THE NEGRO,*" by J.A. Rogers reports that in 1923, Europeans first discovered *"a hitherto unknown Negro race, the Nakhis, 200,000 in number, in Southern China"* King Tang or Ta, the earliest documented rulership of China was the Shang Dynasty [or Chiang] c. 1500-1000 B.C.E., which is credited with bringing together the elements of China's earliest known civilization. The Shang were given the name Na-khi meaning 'Na'-Black and 'Khi'-men or "Serpent Kings" and the 'Snake' or Serpent' later becomes the 'Dragon' or "Sons of the Dragon" The first Chinese emperor, the legendary Fu-Hsi (2953-2838 B.C.) was a woolly haired Melanite. He is credited with establishing government, originating social institutions and cultural inventions. He is said to have originated the *"I CHING"* or "*THE BOOK OF CHANGE*", which is a remnant of the Af-R-Kam 'Ifa' system, called the oldest, and the most revered system of prophecy. Herodotus, the so-called Greek historian stated that there were two Ethiopian (Cushite) people, one in the West; Africa, and one in the East; so-called India, originally called 'Harrappa' from the Ancient Egiptian Neteru [Gods or Forces of Nature] "Haru Ra Ptah," meaning "The Light Rays of the Awakener [Opener]; the Egiptian word Ptah/Putah is the oldest origin of the orient name 'Buddha'. Those in the East were called the Tamil people, the pre-Dravidians, later Dravidians [so-called Sudras]. 'Tam' is the ancient Egiptian word 'Tem' for the "Setting Sun" and 'IL' is 'EL' the Egitptian word 'Neter' and the Phoenician word 'God'. In other words, they were the followers of "Amen Ra," the "Hidden Sun." Now, let's go back to about seven thousand years ago after the destruction of Atlantis, we resettled our empire in Sumeria; the land was called TaMaRe, or Earth, Water and Sun. For the Moors that think they are Black, African, Afro-American, African-American, Latino-American, Puerto Rican, Dominican, Trinidadian, Argentinean, etc., you should know the etymological origin of who you are, cause you're not any of these, they're Moors. Hence, TaMaRe is the nation of Moors; the lands that you call Europe, Spain, Africa, Asia, North America, and South America. After centuries of deception, the term "Moor" has been made synonymous with "Arab" or "Muslim." A simple way of determining who the Moors really were is to first examine the origins of the word, "Moor." Originally from the Greek word, "maurus," it means "blackened" or "charred." From "maurus" we get "Mauretania" the Greek word for the entire continent of Africa, not just sub-Saharan

Africa. From this word, we get modern-day Mauritania, where medieval western Europeans believed to be home to the great West African Moorish civilization. That's right; "Moor" has always been synonymous with "Black African" until recent decades. For evidence, you need only examine the people who named Africans, "Moors"- the Western Europeans. There you'll find countless depictions of Moors as unmistakably Black Africans. Moorish civilization is defined as the Black civilization of Western Africa in present-day Mauritania, Morocco (which is also derived from the word, "Moor"), Mali, and Spain. From this empire came huge castles, colonization in Europe, exploration to the Americas, and sophisticated technology. Centuries before great structures existed in places like France and Germany, the Moors had built huge castles throughout the Sahara and Southern Spain. There were Moors of the Eastern Hemisphere and there were also Moors of the Western hemisphere, oriental and occidental. We as human beings have obligations to each other, and others have obligations to Moors. Moors have a responsibility to the world, as demonstrated during (but not limited to) the era when civilization flourished as a result of their contributions and influences in Europe. Europe was originally inhabited by Moors, and was named Europa after Queen Europa, a Moorish Sister. Europe was a designation point where Caucasians were taken to for a more suitable climate, and to "rope them in", hence Europa. People were named after their accomplishments, and places were named after people, to be recorded in heritage and history, just as the Roman General John Ciprius Afrikana, was named after his conquering and colonizing of Africa, the land that had already been divided by the Atlantic Ocean (Atlantis), along the East Coast which is where Lemuria (*French Le= the, asin the Muurs, or Land of the Muurs*) was/is located. That division is clearly noted on the map, upon looking at South America and Africa, it fits like a puzzle. Now, we also know why the Olmec heads with unequivocally African features, were dug up in South America. Anyone who continues to hold on to the lie that we only came here by way of slave boats is making a clear contradiction to the claim that we are the Mothers and Fathers of Civilization, and that they themselves are agents for the lie. The Caucasoid Mountains were named after the scientist Yaqub KauKasus Kushitus, thus Caucasians. In the mountain caves (as in cave-men, or cave dwellers), they could avoid the rays of the Sun, as many died from skin cancer (leprosy). The on-going education and welfare of the modern man was on the shoulders of those who created them. Moors established universities which to this day offer a pristine education. When the Moors entered or re-entered Spain (711 A.D.), it was for the purpose of civilizing; not conquering. Students flocked from France, Germany and England to drink from the fountain of learning. When Moors left Spain, the Dark Ages ensued, or returned, leaving modern man to fend for themselves. The Dark Ages were not referring to Moors, but those whom the Moors civilized, and gave a basis of education as a matter of obligation and responsibility. Eventually they became known as the Colonist. A colonist is one who essentially leeches off of another, including the land, resources, etc. Unfortunately, due to their desire to have the power and control, and their inability to survive and flourish; as was when the Moors left Spain causing its degrade into beggars, friars, and bandits. Due to their desire to be 'free' and sovereign, they began a mission of conquering and colonizing the world, by any means necessary. Their intent is described in the "Bull Inter Cataera Divina" (a Writ, or Bulletin that describes what they call a divine purpose and intervention, although it was/is not based in divine principles, as it's directives state the following:

> *...To plunder the New World of its treasures was acceptable ecause it was populated by pagans. To Christianize the pagans was necessary because it was part of God's plan; to kill them was right because they were Satan's or Antichrist's warriors ...*
>
> *- Christopher Columbus*

Now, here is where it gets crazy. The slave trade started in America; they took us to Spain and Europe. I know you're like, huh. Yep, it's true. Let me say it again, the slave trade started in America; taking Moors from America back on the north equatorial current to Spain and Europe. It's documented, so look it up. When the Spaniards came to America, they found Indios, which is Spanish for with God, or the people who are with God. That's where the word Indian comes from; Indios. In the 17th century, the European colonist came to America and we made a deal with them for heavy equipment; the reason being is that we couldn't sell our goods because more sophisticated forms of farming were being produced. So, they agreed to give us a certain amount of heavy equipment which would be paid back in six months, if I'm not mistaken. However, they didn't hold true to the treaty because it was oral and not a written agreement. Now instead of them coming back in six months, it was six weeks. When our ancestors couldn't afford to pay them, they took our woman and the girl children from us; and they started having children by them. What happened was within a year's time, we were trying to hold true to our word and yet we found out that these French, British, and Portuguese Europeans were having sex with our woman and girl children; which led to the Yamassee massacre, presently located in Milledgeville, Georgia. Look it up. The Yamassee began to massacre Europeans and push them off their land, but what they had to take back into consideration is that a lot of their woman had been raped and produced mulatto children. Those mulattos grew up wanting to leave the land, presently called reservations; and seek out lighter people. In doing so, they produced these Caucasian looking people that called themselves Yamassee. So after the massacre, a war broke out and we blended in with the Seminole, Creek, Wichita, Seneca, Onondaga, etc. because the Europeans; Dutch, British, Polish, and the French, banned together against us. We migrated, those in Georgia became Seminoles, and others went as far as the Yucatan; we scattered in different directions and we became mixed in with the Cherokees. So you'll find a lot of Yamassee saying "I'm a Cherokee" because their grandmother was a Cherokee, but it goes farther than your grandmother, what about your great-great grandmother? Now, with this in mind, we know that people already were here when Columbus came. Columbus didn't come by himself, it's funny that Moors navigated his ships meaning Columbus didn't know where he was going, but we knew which proves that we've been here prior to their colonization, plus we spoke the dialect. This is not some made up history; I am a scientist, and if anything, a student of life. You have to remember that there were several different slave trades; each organization basis their philosophy or doctrine around one particular slave trade. For example, the Nation of Islam talks about John Hawkins in the 15th century, then you'll find the Moorish Science Temple talking about the 16th century, Sultan city slave trade, and lastly the Ansaars, who talk about the 17th century slave trade out of Sudan; where the people of Zimbabwe, certain Muslims, sold their people into slavery. Now what they don't teach you is that the Indigenous people that were already in America are the parents of all those people from different tribes in Africa, and became what the Arabs call Ifrikaans; those divided into pieces. During the 15th, 16th, and 17th century you had Moors on the shores of America, who were not Muslims, Ansaars, who were Muslims, and slaves. When the colonist came and started making all these rules and regulations, they weren't concerned with the African slave trade; they didn't care about the Moors from Morocco, Mauritania, or Senegal. They were only concerned with our people from Niger, or the Ivory Coast. You see, the colonist were afraid that the Moors that came in were going to get mixed up with the Moors that were in America and the mulattos, so they didn't allow any mulattos to come in; only dark skinned woolly haired Moors from Morocco. When we as Moors came to America, we had documents and treaties saying that Moors are to be treated as equal citizens to every Caucasian in the country; signed by George Washington, the prince of the states. To digress a step, when you hear about a Native American with two long braids and a straight face, that's not a Native American; that's an Indian, overstand? That's an American Indian, but that's not a Native American. The Native Americans were the Olmecs, with their big round heads and big lips; look at the statues. They were the indigenous people of the land. The Chinese came in and mixed with them and produced what you're calling Native Americans today; but they are Indians. The colonist gave them all our rights and now they're walking around calling themselves Cherokee, Choctaw, Seneca, Cheyenne;

meanwhile, we're walking around saying black, nigga, African, Afro-American, African-American, etc. but we as a people allowed ourselves to be treated like property and accepted these branded ids. So what the colonist did to appease the African was he created the Emancipation Proclamation; a Proclamation is not a law; but a public announcement by elected officials. It is not a law. Therefore, the Emancipation Proclamation of 1863 did not set any slaves free. What it did was standardize slavery; The United States being the model for the standardization of slavery, reducing people from their sovereign capacity and forced them to joined nation states. Then they were able to issue statutes, codes, ordinances, and resolutions on them. A statute, as in a state statute, is not a law; it is corporate policy of the corporation. They did every possible thing to us when we started asking for our rights; they created organizations that would stiffen our growth, drugs, propaganda, etc. The first thing that they did was created religion which got our people caught up in things that don't pertain to our own past. Your past has been rewritten for you; have you ever stopped to even question why? Today you need to know what to look for, study etymology and language; the science of words and overstand the science of creation. Language is the expression of a people based on their culture or systematized methodology of learning by extracting from their internal and external environment. English itself was created by Moors. Don't fall for the black or white game, cause black is the primordial divine intelligence from which all things were created; including every atom. Black is holy despite what you've been indoctrinated to think or believe. I know most will argue and say black comes from old English about the 4th century, but you fail to overstand that, the Anglo-Saxons that wrote old English were Moors. I know you're thinking what; Anglo Saxons are Moors, yes. The Stiles family was one of the most ancient families in Germany, the first Anglo-Saxon tribe. The Anglo-Saxons were an empire of aboriginal Moors who were part of the Order of Karest, or what most call Christ. The name Anglo-Saxon is where they get the word angel; angels, an-gaels, and in Arabic injiyl; it is a Gaelic word which is a word from ancient Irish, who were also Moors. Here, let me prove it to you; here are some Anglo-Saxons: The black pope Saint Gregory I (The Great), St. Augustine (1st Arch Bishop of Cantebury), there is also a painting of Pope Gregory by Master Theodoric of Prague, 1364. I almost forgot, St. Bedes; and Icel (king of Mercia which is present day England), also a direct ancestor of King Alfred, who was a Moor. Look up the Anglo-Saxon Chronicles of Bedes. Yes, the Moors ruled England too, all the way up until the 11th century, until William the conqueror, who is in the linage of the Tutor family and the Otter family which are connected to the Bush family line. These are the people that came in and conquered the Moorish empire in England in the 11h century. I'm just trying to give you some information to help clear up a bit of the confusion out there. Investigate the picture of St. Augustine; we as Moors started the Papacy. King Offa Rex is a descendant of Incelbert, and a leader of the Anglo-Saxon clan about 774 A.D. Check out his currency, an Arabic text on an Anglo-Saxon coin. Look it up. Now by 774 A.D. Moors were all over the planet, North America, South America, Europe, Africa, and Asia. You are a documented empire, and that's the truth that's been kept from you for all these years. You're royalty, claim your birthright. You are from the lineage, and offspring of Jesse, Yashua, Isa, Heru; which means that you are the Holy Grail; the real secret of the DiVinci Code is our bloodline. Benjamin Banneker put it on the flag; red and white stripes, seven red, six white, fields of blue with thirteen stars. So let me ask you what's red, white, and blue; your blood. Red blood cells, white blood cells, blue blood cells, deoxygenated blood. To reiterate it again; your blood is red, white, and blue so it is represented in the flag. It's dealing with linage, offspring. We lived by monarchies and divine blood lines. We setup up republics other people to maintain peace amongst themselves, and we regulated them but we never ran them ourselves. It's true, study your history but beware. Some European Social Engineers (and reconstructors of history) have exercised their Demo-intent to distort history and disclaim many of these ancient artifacts; denying their relationships to the Original Ancient Meso-Americans. Attempts have been made to misrepresent these stoneheads and other artifacts, scripts, and instruments as being unrelated to the true forebearers – being the (now mis-named) natural people, branded as negroes, blacks, and coloreds, etc. Truth needs no apology. A few misled Asiatics amongst our own continue to support the seperation tactics initiated by Europeans, and propagate teachings that separate their brother Moors of Old Amexem (Old Mex / Olmecs) from the peoples known as Asiatics/Africans today. The Mayans, the Incas, and the Aztecs of Ancient Central Amexem / Africa / America are anthropologically known as the Nauhautian (mixed) Moors. They too, are descendants from Ancient Moabites / Africans / Asitatics. Other compromised divisionists seek to create divisions amongst their own people. Nevertheless, true world History, and knowledge of pre-Columbian historical evidence consistently proves their claims to be void of validity or truth. Sincere and studious research counters and shows contradictions in their euro-centric claims, and counter to their failures to acknowledge the Ancient Asiatic / Africans / Moabite / Canaanite Progenitors. These same disclaimers admit, on the other hand, that the so-called Canaanites / Africans are the Mothers and Fathers of Civilization......

• **Circa 480 A.D., The Monastic Brotherhood (Catholic Moors from Morocco) landed on present day Connecticut (North America), near the coast of Long Island Sound. The inscription found on granite outcrops in Cockaponset Forest, CN., and the inscription on Haj Minmoun Rock located in Figuig toward the east of Morocco, confirms the voyage. See: Moroccan daily Newspaper (Le Matin D Sahara Et Du Magred, September 16, 1995).**

• **Circa 700-800 A.D., Several Muslim Schools in North America; Valley of Fire (Nevada), Mesa Verde (Colorado), Mimbres Valley (New Mexico) and Tipper Canoe (Indiana). North African Arabic and Old Kufic Arabic scripts are engraved on rocks, test, diagram, charts including writing, reading, arithmetic, religion, history, geography, mathematics, astronomy and sea navigation. For more info, contact: Dr. Barry Fell, at Harvard University.**

- Circa 711 A.D., The Moors that ruled Moslem Spain and Portugal for centuries were black or dark skinned people. See: Golden Age of the Moors, Edited by Ivan Van Sertima Pg 337.

- Circa 1492 A.D., On Monday October 21, 1492, Christopher Columbus admits in his papers, while sailing near Cuba, he saw a mosque on top of a beautiful mountain. The ruins of mosques and minerats with inscriptions of Quranic verses have been found in Cuba, Mexico, Texas and Nevada. The dress of the Indian (Moorish) woman long veils" the men "Breedclothes painted in the style of Moorish draperies" in Grenada and Trinidad... See: Precolumbian Muslims in the Americas by Dr. Yousef Mroueh

- The descendants of these North American Moors are the present day Iroquois, Algonquin, Anasazi, Hohokam, Olmec, Apache, Arawak, Arikana, Chavin, Cherokee, Cree, Hupa, Hopi, Makkah, Mohawak, Naca, Zulu, Zuni... These words also, derive from Arabic and Islamic root origins. See: Precolumbian Muslims in the America by Dr. Yousef Mroueh.

- Circa 1503—1517 A.D., An estimated 3,000 Aborigina-American (Moors) were captured from the eastern seaboard of Terra Nova (North America) some of their names are Ali, Melchor, Miguel, Manne, Juan, Pedro, Antonio and Juan-Amarco. A record of that account can be fond in the Slave Books of Seville, Valencia, Catalina Spain. They were classified as Negro (Negro means Dead) and Black (dirty and evil). See: African and Native American by Jack D. Forbes, page 24.

- Circa 1676 A.D., The Europeans that arrived in New England (North America) described the Aboriginal-Americans (Moors) to be BLACK AS GYPSIES (E-GYPTIANS).

- Circa 1763 A.D., On October 7, 1763, King George R., of Great Britain's, Treaty with the Indigenous People (Indians) regarding land acquisitions and demarcation lines in America. The FOUR Colonies distinct and separate governments are called Quebec, East Florida, West Florida and Grenada. See: Washitaw de Duglahmoundyah Empire, Newspaper, December 1998, Front Page.

• Circa 1774 A.D., The five pointed green star in the center of a field of red is the Moorish flag was the alleged cherry tree that General George Washington, chopped down. See: Moorish Civic Relations Concepts, Volume 14, Page 37.

• Circa 1774 A.D., On October 20, 1774, British-American subjects of the British Empire form the First United Stated of America perpetual Constitutions in the Thirteen Colonies called, "The Articles of Association"'" recognized Moors as Moors not Negroes or Black-A-Moors. See: Journals of the Continental Congress, 75—78.

• Circa 1774 A.D., Noah Webster and his associates branded the Moors Black-a-Moor, Moor was dropped and replaced with the customary term Nigger, Negroe, Colored or black. The use of the word Nigger or Negroe represents the spiritually dead people and not the Nigritian People. Black-A-Moor, n. [For black Moor] A black man or woman, esp. an African negro; any very dark-complexion person. See: New Century Dictionary of the English Language 1927.

• Circa 1775 A.D., The first President of the Untied States of America under the Articles of Confederation was John Hanson, alleged Black-A-Moor, a Maryland Shanwnee Native American patriot who fought in the American Revolution. See: Nuwabic Moors Newspaper, August 7, 1991.

• A Moorish-Mason by the name of Ben Bey Emmanuel Mu Ali a/k/a Benjamin Bannaker, was the architect who designed the streets of Washington, D.C., with masonic codes and astrological glyphs. See: Americas Oldest Secret the Talisman, U.S. Mysterious Street Lines of Washington, D.C. by the Signature of the Invisible Brotherhood. The autobiography of Benjamin Bannaker.

- Circa 1787 A.D., Assisted by England, Scotland, Ireland, Netherlands, France, Germany, Finland and Sweden the United States of America ended their war with the Moors (Moroccan Empire) and signed the Treaty of Peace and Friendship with the Emperor Mohammed III (Moorish-Mason). The aforementioned treaty is the longest unbroken treaty in the history of the United States. See: U.S. Moroccan Relations, by Robert G. Neuman, Former U.S. Ambassador to Morocco (1973--1976).

- Circa 1789 A.D., On December 1, 1789. The Ninth President of the United States George Washington, apologizes to his Masonic Brother Emperor Mohammed III, for not sending the regular advices (tribute: a payment by one ruler or nation to another as acknowledgment of submission or price of protection, excessive tax). Also, President Washington asked the Emperor to recognize their newly formed government.

- The Moroccan Empire (Moors) were the first nation to recognize the thirteen colonies as a sovereign nation. Allegedly the Emperor agreed to their recognition because 25 Moors were members of the first Continental Congress. See: The Writing of George Washington from the Original Manuscript Source 1745—1799, Editor John C. Fitzpatrick, Volume 30, pages 474—476.

- Circa 1790 A.D., On Wednesday, January 20, 1790, A petition was presented to the House of Representatives from the Sundry (numerous) Free Moors, Subjects to the Prince under the Emperor of Morocco in Alliance with the United Stats of America. The Sundry Free Moors Act states that all Free Moors may be tried under the same Laws as the Citizens of (South Carolina) and NOT under the Negro Act. See: South Carolina Department of Archives and History: SC House of Representatives Journal, 1789—90, p. xxii, 353—364, 373—374: In Re. Sundry Free Moors.

- Circa 1857 A.D., The DRED SCOTT Case from the United States Supreme Court; holds that Africans [Moors] imported [captured in an undeclared war of enslavement]. Into this country [Territory of the United States and Several States] and SOLD as [perpetual} Slaves, were not included nor intended to be included under the word "Citizen" in the Constitution, whether emancipated or not, and remained without rights or privileges except such as those which the government might grant them. See: Dred Scott v. Sandford, 60 U.S. (How.) 393, 15L Ed., 691, Blacks Law Dictionary 6th, Edition, Page 495.

- Moors/Africans can not be U.S. Citizens because the Moroccan Empire has a business arrangement with the British Empire [European Corporate Contract Citizens Caucasian Men]. The United States is a foreign European corporation conducting trade and commerce in foreign lands. See: In Re Merrian's Estate, 36 N.Y. 479, Affirmed in U.S. v. Perkins 163 U.S. 625.

- The Hidden History of the Moorish People with the United States of America is recorded on the back of a Federal Reserve Note. There are two seals on the back of the $1.00, Federal Reserve Note (U.S. Currency) on the left side is the Great Seal of the Moorish Empire and on the right is the Seal of the United States. There are over THIRTY THREE (33) passwords on the $1.00 (Note). The INDIGENOUS SOVEREIGN PEOPLE (Moors) were snaked (betrayed) by some of the European Colonial State Citizens who enslaved the Moors and branded them nigger, negroe, black, colored, afro, hispanic, west indian, etc., In order to conceal their true identify. See: Annointed News Journal, June 1998, Page 23. America is the code word for Africa d Morocco is in Africa. See: AmeRICA decoded is AfRICA and MoRoCo decoded is aMeRiCa.

- Circa 1913 A.D., Knowledge of our Moorish Heritage would have been lost if it was not for the Moorish-Mason, our illustrious Brother Noble Drew Ali, who founded the Moorish Science Temple, in Newark, New Jersey (1913). For the unconscious de-nationalized Moors i.e., negroes, blacks and coloreds, Moorish represents our Nationality. Science represents our Ancestors Spiritual Arts maintained in Esoteric Free Masonry, and the Temple represents our Body the dwelling place of the Creator of the Universe. Also, Noble Drew Ali, is responsible for the Moorish flag flying once again on American (Moroccan) soil in 1913. The State of Morocco was not allowed to fly the Moorish flag until 1956 A.D., after their independence from France.

• Circa 1933 A.D., The city of Philadelphia, Pennsylvania, recognize the Moors domiciling in America and their Moorish Titles: El, Bey, Ali, AL, Dey, ect. See: House Resolution No. 75 Legislative Journal (Philadelphia) May 4, 1973, page 5759.

• Prior to Circa 46 BC., the Ancient Moors were referred to by their National names like Washitaw, Almoravides, Almohades, Moabites, Canaanites, Yisraelites, etc... See: Circle Seven Holy Koran.

• The copper colour American Hebrews (Yisraelites) kept the Passover called the Green-Corn Dance. See: History of the American Indians by James Adair (1775) page 80, 101. The Ancient Ones or Mound/Pyramid Builders of North America (over 150 unearthed some Egyptian styled are up and down the Mississippi River) according to the U.S. Bureau of Ethnology was built by copper-hued skin (Asiatics/Moors). See: U.S. Bureau of Ethnology. 12th Annual Report, 1980—1891.

• The Aboriginal Americans or Mound Builders of North America built ceremonial mounds that date as far back as 5,400 years ago. The oldest mound in North America to date is found in Watson brake Louisiana, 32 km South-West of Monroe. See: "Japan Times Newspaper", September 20, 1997, reported by Joe W. Sanders.

• Circa 1848 A.D., On June 6, 1848, a Supreme Court Decision read by Theo H. McCaleb (Judge). Declared that the United States DOES NOT own the land of The Ancient Ones

(Uaxashaktun) Mound Builders of North America (more than 1,000, 000 square miles of land). Also, the Court declared the lawful land owners are the heirs of Henry Turner (Washitaw-Moors/Muurs). See: Case No. 191, U.S. Supreme Court, United States vs. Heirs of Henry Turner

• Circa 1993 A.D., The present day Empress Her Highness Verdiacee Tiara Washitaw Bey, she is the living heir of the Ancient Ones (Empire of Washitaw de Dugdahmoundyah); they are recognized by the United Nations as the oldest people in the world.

Society

It seems that the future of the progeny will be left to fend for themselves without the benefit of knowledge of self. How can one stand as a man or wombman without knowing who you are, and knowing the blood of your ancestors, that continues, and will continue, to run through you, which makes up who you are? It is best advised, that you study to know these things and stop the attempts to discredit our ancestors, thus take your place, amongst the affairs of men and wombmen. We are under occupation, and we have foreign Europeans claiming to be American, which they are not. They have you fooled into thinking you're suffering from prejudice; but actually its theft. You have these leaders out here brainwashing you with that Civil Right stuff which was a fraud struck down by the Supreme Court in 1883 and 1886, which is unconstitutional and has nothing to do with reality or our condition as a people. I deal with knowledge; I don't care what your belief is; that's personal. This is our government, and this is our land. If you don't claim it, the Illuminati will claim it. They've brainwashed you into thinking you owe them, when in actuality, they owe you. They should be charged with kidnapping instead of us trying to become a beneficiary of the fraud; money is only a fiat note. So I suggest to you that you learn to overstand government and what contracts are; then nationalize for your own good. Don't do it because someone says so; you need to study and overstand exactly what it means because the world is changing so fast and you're going to find yourself in trouble. This is the truth, you can't let people fool our people; it's not about what you believe, it's about what you know. Our people don't overstand government because we've been held out of the principles of civilization for so long, and we're not going to get back in it with emotionalism; you're going to have to study. Remember that to have is not been told; if I told you everything, you'd go back to sleep. Be careful, because some of our own people will try to put you back in slavery, so don't think because someone has a little bit of consciousness; that they're honorable. Knowledge and truth alone will set you free. Let me clarify. The Emancipation Proclamation of 1863 did not set any slaves free. What it did was standardize slavery; The United States being the model for the standardization of slavery, reducing people from their sovereign capacity and forced them to join nation states. Then they were able to issue statutes, codes, ordinances, and resolutions on them. A statute, as in a state statute is not a law; it is corporate policy of the corporation that calls itself the Commonwealth of whatever state they represent. A code is not a law; The United States Codes, etc. They are codes, ordinances and resolutions of a municipality of the city you reside, which is a private non-profit corporation. The reason these codes, ordinances, etc. are not law is because the only people who can issue law are people who are acting in their Sovereign capacity, i.e. Moors. These people who sit in these positions as elected officials are not in fact in their sovereign capacity; they're in a corporate-ward status, meaning that they are wards of the state, they are members of the corporation which is a non-profit that calls itself the Commonwealth of whatever state you reside. As long as they have a birth certificate on record; with that birth certificate being a contract. A birth certificate is a contract, and as long as you have a contract with the Commonwealth of your state; you belong to them, and that's what slavery is... Now to further clarify, only people in there sovereign capacity can use law. If you are a member of a corporate-ward state, or corporate-ward nation that calls itself the United States of America; you are a citizen.

A citizen is not a sovereign, a resident is not a sovereign; therefore if you use an address which is a fictitious number associated with a designation issued by a corporate-ward, then you become under the jurisdiction of those people who are also corporate-wards but who are also slaveholders. So if you are operating under that capacity, law doesn't apply to you. If you are a citizen, resident, etc. of the city you reside, which is a private non-profit corporation, then the statutes, codes, ordinances, and resolutions of that private non-profit corporation apply to you. If you are a citizen, resident, etc. of the United States of America, which is a private non-profit corporation then the statutes, codes, ordinances, and resolutions of that private non-profit corporation apply to you. But if you are a sovereign of the Moorish Empire, then those statutes, codes, ordinances, and resolutions of that private non-profit corporation do not apply to you because you are not a member of the corporate-ward state. Its as simple as that. Trust, they understand the difference, and this is why on their documents they use words of art; the word label, the word person, they use the word address; all of these things to try and place you within their jurisdiction, and you unknowingly fill out forms everyday and every time you fill out a form you enter into a contract.

I don't care what kind of form it is; it's a contract. A driver's license application is a contract, a social security application is a contract. When you call up the telephone company and you make a verbal contract over the telephone; this is why they can bill you. When you sign a deed it is a contract, when you fill out a voter registration form, it is a contract; please over stand that. Anything that you put your signature on becomes a contract. The fact that you are not in your sovereign status, means that when you make a contract you were a minor. They don't care; they know you are a minor because to be other than a minor you have to be in your proper person (In Propria Persona) at law. When you are in your corporate-ward status, you look like this to the court (Pro Se), meaning when they get you into court they bring in a prosecutor (pro se cuter). Got that! Now, if your In Propria Persona; the prosecutor cannot come into the courtroom and say anything to you because your not in Pro Se status. The issues of law are threefold:

1. Status
2. Jurisdiction
3. Adjudication

The first thing that happens when you walk into a courtroom is they already make the assumption that you are a ward of the state; and that you don't know any better so they immediately star adjudicating you. As Moors, the first thing you should do when you walk in a court is place your status on the record. Come in with our flag, our Treaty, our Constitution which we signed with them; which is a contract between the United States Republic and the United States of America. There are certain clauses in that contract that apply to Moors which takes us back to the Treaty of Peace and Friendship of 1787, but that contract between the United States Republic and the United States of America does not apply to us other than that part that discusses the Treaty. Its not relevant for us to go into any courtroom and argue about their ordinances, codes, statutes, etc. The only thing your in there to talk about is if we have violated the Treaty; have you violated Article 6 of the constitution? That should be the only question in the courtroom because the status of the individual is between you and the judge; the prosecutor becomes invisible. You say to the judge.. "My honor, what is your status?" "What is your name?" "What is your nationality?" Don't let them get away with saying they're U.S. citizens; you cannot be a citizen and sit a bench, you cannot be a member of the American Bar Association, which is a private non-profit corporation that issues registration numbers to its members who do not have a license to practice law in any state. In order to have a license to practice law you must be In Propria Persona; and in order to be In Propria Persona in this country, whether that be North America, Central America, or South America; you must be a Moor. If anyone attempts to serve you any papers, ask them their status. Where is your license to practice law? Where is your proof of naturalization in my land? Who gave you the authority to be here? Who issued you authority to act in any capacity to file any complaint or charges against anyone? Trust, they can't answer those questions. So, if you walk into a courtroom and nobody has the status to be there; automatically challenge the jurisdiction of the court. There's nothing else that you need to say. There's nothing else that needs to be said! Once you say everything that needs to be said they know everything else to do, but understand they don't want you exposing the truth to their corporate-wards because every slave you take out of their corporation they loose money. This is a business. Understand its not court cause its not law! Then ask yourself, do you understand the law that applies to you or that applies to the matter? There's two types of jurisdiction: the first is jurisdiction over the matter and the second is jurisdiction over the person. Does the court have both? The court must have both in order to proceed; they don't tell you these things but its very important. So, if the court has gotten to the point of adjudication then they are ready to start prosecuting you, or find you guilty of something. The reason for that is because it's a business. They adjudicate you for the sole purpose of obtaining money.

> *"There is no discretion to ignore lack of jurisdiction." Joyce v. U.S. 4742D 215.*

> *"Once jurisdiction is challenged, the court cannot proceed when it clearly appears that the court lacks jurisdiction, the court has no authority to reach merits, but, rather, should dismiss the action." Melo v. US, 505 F2d 1026.*

> *"The burden shifts to the court to prove jurisdiction." Rosemond v. Lambert, 469 F2d 416. "Court must prove on the record, all jurisdiction facts related to the jurisdiction asserted." Lantana v. Hopper, 102 F2d 188; Chicago v. New York, 37 F Supp 150. "A universal principle as old as the law is that a proceedings of a court without jurisdiction are a nullity and its judgment therein without effect either on person or property." Norwood v. Renfield, 34 C 329; Ex parte Giambonini, 49 P. 732. "Jurisdiction is fundamental and a judgment rendered by a court that does not have jurisdiction to hear is void ab initio." In Re Application of Wyatt, 300 P. 132; Re Cavitt, 118 P2d 846. "Thus, where a judicial tribunal has no jurisdiction of the*

subject matter on which it assumes to act, its proceedings are absolutely void in the fullest sense of the term." Dillon v. Dillon, 187 P 27. "*A court has no jurisdiction to determine its own jurisdiction, for a basic issue in any case before a tribunal is its power to act, and a court must have the authority to decide that question in the first instance." Rescue Army v. Municipal Court of Los Angeles, 171 P2d 8; 331 US 549, 91 L. ed. 1666, 67 S.Ct. 1409.* "*A departure by a court from those recognized and established requirements of law, however close apparent adherence to mere form in method of procedure, which has the effect of depriving one of a constitutional right, is an excess of jurisdiction." Wuest v. Wuest, 127 P2d 934, 937.* "*Where a court failed to observe safeguards, it amounts to denial of due process of law, court is deprived of juris." Merritt v. Hunter, C.A. Kansas 170 F2d 739.* "*the fact that the petitioner was released on a promise to appear before a magistrate for an arraignment, that fact is circumstance to be considered in determining whether in first instance there was a probable cause for the arrest." Monroe v. Papa, DC, Ill. 1963, 221 F Supp 685.*

"*An action by Department of Motor Vehicles, whether directly or through a court sitting administratively as the hearing officer, must be clearly defined in the statute before it has subject matter jurisdiction, without such jurisdiction of the licensee, all acts of the agency, by its employees, agents, hearing officers, are null and void." Doolan v. Carr, 125 US 618; City v Pearson, 181 Cal. 640.*

"*Agency, or party sitting for the agency, (which would be the magistrate of a municipal court) has no authority to enforce as to any licensee unless he is acting for compensation. Such an act is highly penal in nature, and should not be construed to include *anything which is not embraced within its terms. (Where) there is no charge within a complaint that the accused was employed for compensation to do the act complained of, or that the act constituted part of a contract." Schomig v. Kaiser, 189 Cal 596.*

"When acting to enforce a statute and its subsequent amendments to the present date, the judge of the municipal court is acting as an administrative officer and not in a judicial capacity; courts in administering or enforcing statutes do not act judicially, but merely ministerially". Thompson v. Smith, 154 SE 583.

"A judge ceases to sit as a judicial officer because the governing principle of administrative law provides that courts are prohibited from substituting their evidence, testimony, record, arguments, and rationale for that of the agency. Additionally, courts are prohibited from substituting their judgment for that of the agency. Courts in administrative issues are prohibited from even listening to or hearing arguments, presentation, or rational." ASIS v. US, 568 F2d 284.

So when any government agency, official or subsidiary says a person must, a person shall, etc.; if you think that means you, then you're wrong, but it's presented to you that way so that you can be tricked into making a contractual agreement saying you're negro or a corporate person. Once you sign the contract, the contract becomes the law, so they sue you on that contract. You need to know that you are the law; an indigenous sovereign in any governmental structure. There are two types of persons in law; one is the artificial person, which is corporate; and the other is the natural person, which is the natural biological breathing thinking entity that you call yourself. It is your bit of sovereignty given in allegiance to that said government that accumulates as the God power that a government executes relative to its regulations of the citizens in a corporate capacity, meaning that they cannot even hold a court over the general citizenry, only the general citizenry can. The governments power is limited to the judicial branch of that government, overstand? Licenses comes under legislative, and statutory codes which cannot be executed in a judicial court, overstand? The government has been running statutory courts under the authority of judicial powers which is a constitutional violation, overstand? Licenses comes under that crossover, understand? Therefore, as an example, if you were a foreigner and you were operating in these territories or any other government territories under corporate capacity for the benefit of your corporate operations of capital gains, excise taxation regulates that. Under that auspicious, licenses are given by officers of excise. That authority within government is limited to those persons and checked by the constitution dealing with manufactured goods and the movements of those goods; that's were license comes in, but it does not apply to the common people or to their conveyances, overstand? They did that out of the Christian codes of 1724, adopted it into all the states in order to stagnate our movement, i.e. our economy. In order to protect themselves from international lawsuits they wrote your names in all capital letters because it does mean corporation versus non capital letters which is an indication of the natural person. The deal is when they pass that instrument to you, whatever it may be, and you don't object and agree that that's you; they use that contract against you, overstand? The driver's license is not an identity card, it's an excise tax payers card for corporate activity and the benefit derived from corporate activity but because they wrote the name in all capital letters they are protecting themselves legally and lawfully by saying that they were communicating with the corporation and you; if you don't know that and accept it then you agree that you're a corporation and they regulate you accordingly. I hate to say it, but we've been fooled. There is no such thing as the American President, an American Flag; nor American Nation. If so, they'd have to admit that they are foreign corporations, or plantations incorporated; holding the welfare, resources and the properties of the Moorish nation; which is the United States Republic i.e. The

Great Seal which is us. They cannot back up the lies, and they will be challenged by the rest of the civilized world. It's important for them to maintain this racism, separatism, etc; they encourage leaders to keep separating us from each other; it's all a plot because it hides the legal process of them stealing your birthrights. That's the only way it can be legally sustained; but they are in violation of international law, as well as the constitution and can be sued by individuals. I'm sure that's why Johnnie Cochran is no longer with us. They must hide the fact that they use the excise tax as an identity when it really isn't; do you see the fraud in it? You know fraud has a statute of limitation, right? So that means if Moors got together they could have a class action lawsuit and take it to court without having to pay a dime. You have to overstand how the game is played. You have to learn some of the processes in order to use it against them; make it cost them to abuse us. Now, I have to say that most of our people love rituals, but lack substance. Aboriginal people have always governed themselves from the world based on cosmology, which is the study of the origin of life itself, the functional dynamics of the creation of the universe, its operative principles, what power sustains it, and how we as a nation of people should be living in the universe. Furthermore, government is an extraction and a social science that is applied based on an overstanding of cosmology. If you continue to just understand, you will remain ignorant, which is ignoring the facts. Most unconscious Moors continue to claim to be descendants of negros, coloreds, blacks, African-Americans, Puerto Ricans, Dominicans, Latinos, etc. In truth it places them outside of the law, outside of the constitutional fold of government, outside of the ancient principles they say reside within them; outside of the human family. They are outlaws and don't seem to recognize those branded names are not withstanding, as well they don't seem to acknowledge that they are in dishonor of their ancestors. It is time to stop claiming those brands. When it comes to society I will attempt to show you that you are undeniably Moor; and at war. Ready, let's go. We have become ignorant by our own actions and it only takes admitting and then putting the things together that we need to do it. After the colonization of our people, we were forced to accept the Christian codes of 1883; to practice monogamy. Presently, we don't practice monogamy, we practice an irresponsible form of polygamy; both based on the fact that you have accepted Christianity and Jesus Christ as your lord and savior and redeemer. If you're a Christian, don't get upset with me but Christian is synonymous with Romans and Greeks. Know that long before there was a Christ, there was a Karest and the Moors are the founding fathers and mothers of the Order of Karest; which is the original doctrine of Christianity which has nothing to do with a Jewish boy, so get mad! To continue, monogamy was forced upon all aboriginal people as a scientific and military activity to induce the ability of capitalism to exist. It makes the Moorish woman a queen, which is a harlot; don't take my word for it, look it up. In that light, our Moorish woman was no longer cared for since she is the dominant number amongst our aboriginal population; without our wombman there could be no life. She is the time machine into this dimension. We have forged ourselves into a family structure that has nothing to do with preserving the family, but it has to do with preserving the capitalism and the ulterior consciousness which we call Europeans; hybrids, or albians. We need to study the family structure of the original and aboriginal cultures. It is imperative. We as a nation of people need to plan to govern ourselves based on cosmology. It was Christianity that stepped in and forced their destructive way of life on our people. They tell you Church & State are separate, ha! It's time to wake up. This has nothing to do with feelings, we are discussing reality. The family structure is the most scientific and loving on the planet. Oh, all you men juggling woman and playing with their feelings, you're disrespecting the mothers of civilization and enunciating chastisement on yourself because you're not a mathematician; you're a sucker. Same thing for you Sunday worshippers; church on Sunday, strip club on Friday. This is real. Do you want to stand up and be yourself, or do you just want to pretend; stroking your own validity based on belief and ignorance? We as brothers and sisters, Gods and Goddess must begin to teach our children the original family structure of the aboriginal people. Our structure of family is the mathematics of nature, thus the symbiotic relationship between man and wombman and the culture of our people. I know most of you think you've been free, but you're still slaves; arrogant, and not willing to say we as a whole have messed up, we've got to clean this place up, go green, etc. The world knows who we are and they've been waiting for us; everything we do they try to emulate, right? We need to take action towards change. Ok, we messed up, now let's rectify it. We're the Gods and Goddesses of the universe. We are!! I don't say that to make you feel good, it's written to illustrate your responsibility, and it is an actual fact. Most believe in a mystery god, which has nothing to do with sacred geometry, science, mathematics, cosmology; you were taught to not study and just believe another man's word, a foreigner at that. You were taught not to question, or else. We as a collective group of minds need to start asking some questions. Now to get back on track, cosmology rules Gods and Goddesses. In order for us to get our nationality and birthrights we must properly know our history and culture. God is in your eye, not in the sky. Most people are walking around fighting each other; we are all one people, what is wrong with us? Love each other and show compassion, damn. Why is this most difficult? We are in the age of Aquarius, as of 1998. The old Piscean leaders who teach you mysteries are dead; they cannot build, they cannot do anything. What are the words that can come out of the word Aquarius or Aquarian; Quran, Ausar, Aryan, iron, Iraq. It's telling you what's going on. The Aquarian Age is a fixed air sign which means something will be firmly established. An air sign deals with mental abilities and communication; so something will be firmly established in the mental, or mentality. So Saturn and Uranus will rule the mental place; this is cosmology telling you the rhythm of the time. Saturn represents the law; Uranus represents revolution. The reason why these planets represent what they do is because they are made of certain minerals and elements, and as they move in their orbital frequency they impact human consciousness. So to reiterate, we are dealing with the Aquarian Age which is the unveiling of secrets; you're dealing with revolution in law. The age is calling for a revolution and overstanding of law which will be firmly established through a heightened level of mental consciousness coming through our communication which deals historically with knowledge of the Moors. What happened in Sumer is how we got into the state our people are in now. Sumer is the actual place where the genetic hybridization took place; not in Yucatan, Mexico. The only way to resurrect our empire is to return to absolute law of the aboriginal people. That's the only way; even if it's uncomfortable. All of this relates to your society as an aboriginal. How this ties in with America is honestly quick simple. Let me explain; the United States of America is in fact, an indebted nation to the Moorish Empire, and their Treaty of Marrakesh, 1787 illustrates their payment of gold and silver. The United States of America has been part of the Moorish Empire until its decline in 1885, at the Berlin Conference and subsequently until 1782 were the formal capitulation papers were signed by George Washington. It was communicated to him that there is no power proceeding from God, but God, and refers to him (George Washington) president of America and prince to all the states. Those states were under the jurisdiction of the Moroccan Empire. George Washington worshiped Allah and confirmed the empire. That's what masons do; they practice the science of Islam; I-Self-Lord-Am-Master. Anyways, there is no escaping Moors responsibility, thus the Treaty of Peace and Friendship is one agreement that outlines some of those responsibilities, and is to be attached to the Constitution as valid against the Constitution, as is the Confederation *(See Article VI of the Constitution for the United States of North America, Republic)*. The Constitution is an ongoing agreement to prior agreements. Many think the Constitution voids any previous agreements. If that were the case Article VI would not have been written, and those who sit in the seats of government right now wouldn't take an oath to it, as it is where anyone who sits in the seats of government derives their authority from. As for the Confederation, if you trace that back you will see, it too is based in divine principles, and the Confederation was those of the aboriginal people, wherein Confederations include a group or conglomeration of people, or a political jurisdiction of people. The Treaty had to be written as a matter of responsibility for those who came here causing discord. Some of us Moors did not want to treat with them, and we were at odds with each other. That is like your brothers and sisters disagreeing, and doing something against your preference. This does not mean they are not still your brother or your sister. The Treaty of Peace and Friendship, surely exist although denied not only by the occupying European Colonists, for obvious usurping reasons, but is also denied by the descendants of those who authorized it, which is one of the greater points. The Treaty(Constitution)was established after the Colonist came here and committed murder, mayhem and treason in this, the land of "Milk and Honey" (Northwest Amexem/Northwest Africa /North America). Take note that the treaty was written after the "Declaration of Independence of 1776, wherein the colonists were determined to be treated equally; because they were not being treated equally. It is also interesting to note how today, they try to act as if they never heard of Moors, yet the Treaty was written, and is documented in American History, as is the letter from George Washington to Muhammed Ibn Abdullah-Sultan of

Morocco, December 1, 1789. The Treaty was written to bring about peace, and to provide protection to them (the Colonist) on the high seas, and on the Land. Thus, the banner they were to hail for acknowledgment was/is the *"Star Spangled Banner"*, just as most of us were taught via the song, in school. It is the Banner of Peace and Amity, or of commerce, which allowed them (Colonist) to do trade/commerce. If we don't stand up, or speak out, others who have an obligaion to us will continue, with no rebuttals, no recourse or remedy. For instance, the Battle of Gettysburg is one of the bloodiest battles fought here, and the graves are filled with Moors (take a trip there and see for yourself). The irony of the south fighting the North for freedom, or not, is that the battle was fought on the real property of what they call a free slave in their modern documentaries. The tables seem to be turned at this time, yet, the pendulum always swings back the other way, and "cause and effect" is always in effect. Moors obligation to others, and others obligation to Moors, is outlined in the Treaty of Peace and Friendship. The Treaty and the Constitution for the United States of North America Republic are equally valid today. One cannot be acknowledged without the other. The Treaty is written perpetually, but is to be revisited every 50 years for any adjustments, or amendments. It was revisited in 1836, the only change was those who were present, as the Sultan had passed form, and his son was present on his behalf. It has not been revisited since then, as those who would treat of it, believe by way of false teachings, that they have nothing to do with it, and that they are descendants of negros, coloreds, blacks, Afro-Americans, African Americans, Puerto Ricans, Latinos, Dominicans, etc. The European Colonist, who needed to preserve the treaty for their own safety and protection, can now bury it and pretend it doesn't exist. They act and speak for Moors because those brand names are chattel property, and fictitious corporate entities that have no rights to secure. This is why in court they want you to have someone else speak for you, and then they sentence you, which is just a phrase of words. Those who think they are negro, colored, black, Afro-American, African-American, etc., and victims of slavery, usually begin their history on slave boats from Africa. Subsequently they also end their history there as well. As soon as we investigate past the 17th century, prior to those brands; we enter immediately into Moorish History all over the World. It is a necessity to establish rules of law, particularly for the uncivilized for the purposes of maintaining civilization. These ancient laws are derived from the matriarchal, natural law, for commonality of all. This is why the Constitution on one hand, is referred to as one of the oldest and ancient documents, and on the other hand, people think it came about only for and by the European Colonist, ratified in 1791. The principles of civilization are ancient, and many have stood for the preservation of them. Downright murder and theft holds no candle to morals, ethics and truth! The destruction of the libraries, burning of the books, has left a seemingly clean slate to introduce mis-information, and that is exactly what happened. The Reconstruction Era is a more recent demonstration of re-constructing the literature, and preparing for a mis-education. This is why one will find that older writings have the greatest appeal. The principles of civilization are ancient, and thus "We the People of the United States", of which there are many, adopted this Constitution for the United States of America, as stated in the Preamble. Everyone is endowed by their Creator, with certain unalienable birthrights, which cannot be bought, sold, or transferred, nor can the exercising of them be made void, or turned into a crime. The Constitution protects preserves, and secures those unalienable rights, and, guarantees a republican (matriarchal) form of government. Anything outside of that has no standing. It is a travesty when the civilization principles which established the rules of engagement are not honored by those who say they are the Mothers and Fathers of Civilization, as they are the very descendents of the ones who established and authorized them by way of writing and securing these principles as the supreme law of the land. Failure to acknowledge and honor your own unalienable rights allows for others to violate them, and has caused great degradation to man and mankind. This continued dishonor can only be rectified by the originators, the aboriginal and indigenous people of the planet, who have a responsibility and obligation! Knowing that the Aboriginal people of Europe, of the planet for that matter, are Moors, will help one to gain understanding as to why Moors claim "free white persons", from a lawful standpoint, as it is not an identification from a pedigree standpoint. It is a legal status terminology. Hopefully it will put an end to any confusion regarding "free white persons". The status of the descendants of the Moorish Inhabitants of Spain and Portugal on American soil is FREE WHITE PERSONS (natural men and women). This status does not apply to the Caucasian Race, Aryan Race, or Indo-European Races under the Naturalization Act (Amended by Act. July 14, 1879), I Stat.103,c.3 See: Black's Law Dictionary (Fourth Ed. P. 797). Now take a look at the Treaty of Peace and Friendship, Article 6, 11, 21, and thus there is no mistake as to the parties involved, and their responsibilities and obligations, and the actions to be taken upon their encounters. The word Moslem/Muslim is synonymous with Moor. This is one of the important things to know in doing your research. They will, in fact, use the word Moor, and then in the same, sometimes sentence, certainly paragraph, they will use Moslem, just as they use the word Christian and Crusader, which are also the same. This is true and indicative throughout older writings, and it is most necessary to know, along with many other names for Moors, such as Ottomons, Turks, Almoravhides, Almohades, Saracenes, just to name a few. Moors Heritage and history goes back as far as but not limited to Freising, Germany. The Moors head appears on the Pope's Coat of Arms today! As well, it is cut into the lawn at St. Peter's Basilica in Rome. Anytime you see in the writings possibly ancient, you can be assured there may not be much said because the modern man, integrated into the human bouquet, is not ancient, thus he either doesn't know, or doesn't want to tell, because he may not want you to know, as has been in the past. However, in the now, the truth of information is flowing freely because this is the unstoppable age of information, or age of knowing. The energy which is described to you in the sign of Aquarius, the humanitarian, of which the time is now to bring the people back into the human family, and concerns of humanity are abound. One can resist this, however upon a real challenge, which means study; those who try to prove facts wrong will find that they are proving them to be correct. Their personal acceptance is based on their willingness to take responsibility, and recognize their obligation.

The Universe

Does life exist elsewhere in the universe? Our galaxy is composed of roughly 100 billion stars. How many of these stars have planets? Scientist are finding that quite a few do. They already know of over 100 stars aside from our sun that have at least one planet. The numbers are: 146 planetary systems; 170 planets; 18 multiple-planet systems (as many as 4 in some systems). At present, the scientist can only detect large Jupiter-like planets in orbits that are relatively close to their stars: these produce the strongest tugs on the parent stars, which is what they measure. It appears that a substantial fraction of stars—at least 5% have planets. As sciences detection techniques improve, they may find many smaller planets and find that indeed most stars have planets. Right now, scientist would not be able to detect our own set of planets around another star using their current techniques; yet we know we are here. This means solar systems much like our own are still beyond our detection limits. Putting the number of stars in our galaxy together with an estimate that ten percent of stars have earth-sized planets, we get that 10 billion stars in our galaxy alone harbor earth-like planets. By earth-like, I just mean rocky planets with masses comparable to earth's mass. Not all of these earth-like planets will be in the habitable zone where we see life in our solar system. For example, Venus is too hot, Mars is too cold, though both are "earth-like" by the definition above. Let's say only 1% of these earth like planets happen to be in the just-right zone. Now we have as many as 100 million habitable planets in our galaxy. They don't know yet how rare life is. With only one planet to guide them, the estimates cover a huge range. But let's say for the sake of argument that the chances for life to form are a remote one-in-a-million given a habitable planet. Some would argue that it's closer to near certainty that life; we're talking single-cell organisms, forms. But with the pessimistic long-shot odds of one in a million, that still gives 100 instances in our galaxy. There are approximately 180 billion galaxies within our 13.7-billion-light year horizon. So now we have 10 trillion instances of life in our universe given the harsh one-in-a-million odds. But wait, there's more. Our visible universe is but a small portion of the entire universe. We know the universe is at least fifty times the size of the visible universe within our horizon. But this is in linear size—radius, or diameter. In volume, the universe is then at least 125,000 times larger in volume than our visible volume. Assuming physics looks the same outside our horizon, there are now about a quintillion instances of life in the entire universe, as a lower limit. The sheer size of our universe and the resulting number of stars and planets is so absolutely staggering so as to overcome the long odds for developing life. One other aspect we haven't reflected on is the enormity of time over which life has to develop. They can't really easily grasp time periods longer than maybe 1,000 years. Yet it takes a thousand of these periods to constitute one million years, which is still short on geological timescales. It would take a thousand of these one-million-year periods to make one billion years. The universe is seventy-six trillion years old, and they found fossil evidence of simple life on the 4.5-billion-year-old earth as far back as about 3.5 billion years. Interestingly, the earth was not very hospitable when it was young, only in the seas. It may have taken only a few hundred million years for life to form once the earth was a calm and hospitable neighborhood. Even this geologically short period is so long that it is truly impossible for their brains to take it in —much like comprehending the vast size of the universe. According to cosmic theory, the universe began with a singular explosion followed by a burst of inflationary expansion. Following inflation, the universe cooled, passing through a series of phase transitions and allowing the formation of stars, galaxies and life on earth. After the so-called big bang exploded with enormous energy, the universe began the cool-down period that has lasted until our time. The resulting chain of events is a cosmic drama with many acts, dramatic transitions and a host of factors appearing and disappearing along the way. The early scenes played out at unimaginable temperatures and densities, the stage set by the fundamental properties of particle physics. These processes had to be finely tuned to yield a universe capable of forming the galaxies, stars and planets we observe today. To reconstruct the cosmic story, telescopes and space probes detect the relics from the early universe, and particle accelerators recreate and study the extreme physics that characterized the stages of development and the transitions between them. As scientist began to understand the cosmic past, they look to the future of the Universe and try to predict its ultimate fate, in a sense; of course it's deeper than that. I thought I would give you a bit of information so we can start getting into your mind and ask some questions. Hopefully, I've gotten your attention. Now, physicists have developed a commanding knowledge of the particles and forces that characterize the ordinary matter around us. At the same time, astrophysical and cosmological space observations have revealed that this picture of the universe is incomplete—that ninety-five percent of the cosmos is not made of ordinary matter, but of a mysterious something else: dark matter and dark energy. Recent measurements with telescopes and space probes have shown that a mysterious force—a dark energy—fills the vacuum of empty space, accelerating the universe's expansion; but they don't know what dark energy is, or why it exists. More is unknown than is known. They know how much dark energy there is because of how it affects the Universe's expansion. Other than that, it is a complete mystery, but it is an important mystery. It turns out that roughly seventy percent of the Universe is dark energy. Dark matter makes up about twenty-five percent. The rest - everything on Earth, everything ever observed with all of their instruments, all normal matter - adds up to less than five percent of the Universe. One explanation for dark energy is that it is a property of space. Albert Einstein was noted as the first person to realize that empty space is not nothing. Space has amazing properties, many of which are just beginning to be understood. The first property that Einstein discovered is that it is possible for more space to come into existence. Then one version of Einstein's gravity theory, the version that contains a cosmological constant, makes a second prediction: "empty space" can possess its own energy. Because this energy is a property of space itself, it would not be diluted as space expands. As more space comes into existence, more of this energy-of-space would appear. As a result, this form of energy would cause the Universe to expand faster and faster. Unfortunately, they don't understand why the cosmological constant should even be there, much less why it would have exactly the right value to cause the observed acceleration of the Universe. Since Einstein, physicists have sought a unified theory to explain all the fundamental forces and particles in the universe. The result is a stunningly successful theory that reduces the complexity of microscopic physics to a set of concise laws. The quest to discover the fundamental laws of nature has led to the revelation that the laws of physics, and the particles they govern, exist because of underlying symmetries of nature; some of them lost since the big bang. One such lost symmetry might be super symmetry. Just as for every particle there exist an antiparticle; super symmetry predicts that for every known particle there also exist a super partner particle. Part of the strong theoretical appeal of super symmetry, an essential part of string theory, is its possible connection to dark energy and the fact that it provides a natural candidate for dark matter, the neutralino. The revolutionary concept of string theory is a bold realization of Einstein's dream of an ultimate explanation for everything from the tiniest quanta of particle physics to the cosmos itself. String theory unifies physics by producing all known forces and particles as different vibrations of a single substance called superstrings. String theory brings quantum consistency to physics with an elegant mathematical construct that appears to be unique. The strings themselves are probably too tiny to observe directly, but string theory makes a number of testable predictions. It implies super symmetry and predicts seven undiscovered dimensions of space, dimensions that would give rise to much of the mysterious complexity of particle physics. Testing the validity of string theory requires searching for the extra dimensions and exploring their properties. At the most fundamental level, particles and forces may converge, either through hidden principles like grand unification, or through radical physics like superstrings. We already know that remarkably similar mathematical laws and principles describe all the known forces except gravity. Perhaps all forces are different manifestations of a single grand unified force, a force that would relate quarks to leptons and predict new ways of converting one kind of particle into another. Such a force might eventually make protons decay, rendering ordinary matter unstable. Physicists have identified 57 distinct species of elementary particles; if I'm not mistaken, and have determined many of their properties in exquisite detail. Perhaps these particles are just different notes on a single superstring. Perhaps they are related by grand unification, or other hidden symmetries, in ways that they have yet to decipher. Unification may provide the key, the simple principle that gives particles their complex identities. They have discovered three families of quarks and leptons, families of fundamental particles that differ only in their masses, which range from less than a millionth of the mass of an electron to the mass of an atom of gold. Just as quantum mechanics led to an understanding of the organization of the periodic table, they look to new theories to explain the patterns of elementary particles. Most of the matter in the universe is dark. . It holds the universe together. Without dark matter, galaxies and stars would not have formed and life would not exist. Recent observations of the effect of dark matter on the structure of the universe have shown that it is unlike any form of matter that they have discovered or measured in the laboratory. The theory of super symmetry predicts new families of particles interacting very weakly with ordinary matter. The lightest super symmetric particle could well be the elusive dark matter particle. Ubiquitous, elusive and full of surprises, neutrinos are the most mysterious of the known particles in the universe. They interact so weakly with other particles that trillions of them pass through our bodies each second without leaving a trace. The sun shines brightly in neutrinos, produced in the internal fusion reactions that power the sun. These reactions produce neutrinos of only one kind, but they mysteriously morph into two other kinds on their way to earth. Neutrinos have mass, but the heaviest neutrino is at least a million times lighter than the lightest charged particle. The existence of the neutrino's tiny nonzero mass raises the possibility that neutrinos get their masses from unknown physics, perhaps related to unification. Detailed studies of the properties of neutrinos—their masses, how they change from one kind to another, and whether neutrinos are their own antiparticles—will tell them whether neutrinos conform to the patterns of ordinary matter or whether they are leading us toward the discovery of new phenomena.

Yin and Yang

"The One gave birth to two things..." Tao The Ching.

In cosmology, Yin and Yang are two opposite but complementary principles that regulate the functioning of the cosmos. Their repeated alternation provides the energy necessary for the Universe to sustain itself, and their continuous joining and separation is at the origin of the rise and the disappearance of the entities and phenomena that exist within the world of the ten thousand things. According to a celebrated statement, which is found in one of the appendixes to the *Book of Change(Yijing)*, "one Yin and one Yang, this is the Tao." This sentence refers to the Tao that first determines itself as the One (or Oneness), and then through the One gives birth to the two complementary principles. As each of these stages generates the next one, Yin and Yang are ultimately contained within the Tao itself. At the same time, the phrase "one Yin and one Yang, this is the Tao" refers to the continuous alternation of Yin and Yang within the Universe. When one of the two principles prevails, the other yields, but once one of them has reached the height of its development, it begins to recede; in that very moment, the other principle begins its ascent. This mode of operation is especially visible in the time cycles of the day (alternation of daytime and nighttime) and of the year (alternation of the four seasons). The complementary couplings of Yin and Yang pervade the entire Universe, and their elemental opposition provides the dynamic tension required for all movement and change. The principles of yin and yang apply to everything from the movement of stars and planets to the most minute cellular functions of the body. Yin is the shady side of a hill. It represents darkness and passivity, and is associated with the qualities of yielding, softness and contraction. Its primary symbols are woman, water, and earth. Yang means the sunny side of the hill, represents light and activity, is associated with resistance, hardness and expansion, moves naturally upward and outward, and is symbolized by man, fire, and heaven. Yin and Yang are mutually interdependent, constantly interactive, and potentially interchangeable forces. Despite their polarity, each contains the embryonic seed of the other within itself, as illustrated by the yin/yang circle. The circle itself represents the ultimate source, each with a dot of its own opposite growing inside it. The ceaseless intermingling of yin and yang gives form to all things. The sexual union of male and female gives life to all things, and thus regarded as the essential earthly manifestation of the great cosmic dance of yin and yang. According to the *Tao Teh Ching*, any force, object or idea is incomplete and meaningless without reference to its own essential opposite. Good has no meaning without evil to define it, beauty is invisible without ugliness to contrast it. The dualism of western philosophy has taught the individual that opposites are split into two mutually exclusive domains, with value attached only to the good, the beautiful and the true. The evil, ugly and false aspects of life are either deliberately and vainly suppressed or simply avoided as bad. Western scientific thought perceives the universe as a scattered collection of static, unrelated objects, each of which has value and behaves according to absolute laws. By contrast, looking at life as it is, one perceives the Universe as a living ocean of flowing forces in a state of constant flux, like waves on the sea, with everything interrelated, interactive and dependent for existence on its own opposite quality. Recognize and balance the opposing forces underlying all situations and phenomena, but no illusions are to be harbored about defeating any cosmic force, positive or negative. Nothing exist or functions except in direct conjunction with its own essential opposite, and all abnormal phenomena are caused by a critical imbalance between these two primordial forces. In nuclear physics, you'll find Yin and Yang at work on the molecular level, for it is the push and pull of opposite forces that literally glues atomic particles together to form matter. But when those molecules are further taken apart, scientist discovered that they cease to exist altogether as matter. Instead, they become vibrating bundles of pure energy organized in certain patterns that give the illusion of solidity when viewed from a distance. Thus establishes as scientific fact that matter and energy are interchangeable. Together, Yin and Yang each prevents the other from distorting things in its own particular light, view, or perception.

The Treasures of Life

"The One gave birth to two things, then to three things..." Tao The Ching

These three things are the treasures of life. The three treasures of life are: essence, energy, and spirit. Unlike the energies of Yin and Yang, which permeates both inanimate and animate worlds, essence, energy, and spirit are associated specifically with human life. Like Yin and Yang, the three treasures of life are distinctly different yet entirely interdependent elements. They comprise the three fundamental levels of existence for all living beings: physical, energetic, and mental. The natural legacy of life conferred upon all living beings at birth. The first treasure of life is essence. There are three basic forms of essence manufactured within the body. The first is blood-essence, which includes all the various vital elements carried into the bloodstream, such as red and white blood cells and nutrients absorbed from digested food in the small intestine. The second is hormone-essence, which comes in two forms: life-essence and semen-essence. Life-essence includes all the vital hormones secreted by various glands throughout the body's endocrine system, and which serves as master regulators for growth, metabolism, sexuality, immunity, aging, and so forth. Semen-essence refers to sperm and related male hormones in men, the ova and related female hormones in women. The third form of essence includes the heavy fluids such as lymph and lubricants surrounding joints and other connective tissue (synovial fluid), as well as tears, perspiration and urine. These are primarily involved in excretion of waste products and dissipation of stagnate energy. Essence and energy are intimately related: 'Qi is the general of the blood; if qi moves, then blood moves.' Immunity and resistance are natural factors manufactured within the body in the form of highly refined essence. Even more remarkable is the scientifically established fact that interferon is produced in the body by only three types of cells; cells which correspond precisely to the three forms of essence which provide the key factors of immunity. The first source of interferon is leukocytes, or white blood cells. This corresponds to the blood-essence. The second sort of cell that has been proven to produce interferon is fibroblasts, which are special cells associated with connective tissue; this correspond with the fluids associated with joints and connective tissue as the second source of immunity factors. The third source of interferon is a cell called T-lymphocytes, or T-cells more commonly known, which produce a variety of interferon that works directly with the DNA master-molecules in transcribing genetic messages and is therefore associated with sexual fluids. Hormone-essence , especially sexual hormones, are the third major source of immunity factors. The second treasure of life is energy. All forms of life in the universe are animated by an essential life force called qi, ether, energy. Qi literally means breath, air, energy and is the exact equivalent to the term prana in Indian philosophy. The nature and connection between the forces of the outer universe and the forces of the human body are developed, structured, and influenced by groups of stars, including the planets and cosmic particles within its configuration. In addition to energy from the stars, humans require a source of natural energy supplied by the earth. Absorption of this all natural force nourishes the nervous system, organs, glands, senses, etc. There are three classification of this energy, or force. The first is the original force, or heavenly energy which manifest as the energy of the stars, planets, and galaxies. This all-pervading force nourishes the mind, soul, and spirit of each individual. This force is drawn to our planets due to the relationship between the earth and moon; combining to create a strong magnetic pull that attracts and then sends to earth the energies of the stars in our galaxy. The second force is the cosmic particle force; a result from the exploding stars that die and drift through space as fine particles. The magnetic power of the earth and moon attract many of these particles; they then drift through earth's atmosphere as dust, thus eventually becoming soil. Thirdly, the force of nature, or earth force includes the energy of plants, animals, water, and all the natural occurrences of the planet earth. The plants and trees extend themselves to the sun, stars, and cosmic particles using it for sustenance and growth, allowing animals to consume the vegetation; profiting from the cosmic energy. Qi is invisible, silent, and formless - yet it permeates everything. Qi takes many forms within the human system. The most basic form is called primordial energy. This refers to the original burst of pure energy that occurs at conception and breathes life into the fetus in the womb. It begins to dissipate from the moment of birth, and the rate of dissipation determines one's lifespan. By drawing on these reserves of energy to compensate for poor diet and other bad habits one accelerates the rate of energy dissipation and sow the seeds of chronic debility in adulthood. Primordial energy may be tonified and enhanced through diet, herbs, proper breathing, regulated sex and other disciplines (tai chi, qigong, pranayama) aimed at recharging primordial energy, retarding the rate of dissipation and thereby prolonging life. Another form of qi is called vital energy in its volatile, kinetic, active form. It is the sort of energy that builds in the body during the excitement of sexual intercourse and is released in a burst during orgasm. It is like the energy drawn from a battery in the form of electric current. It is absorbed directly from the atmosphere when breathing. The body produces two distinct forms of qi directly from the essential nutrients extracted by digestion from food and water. One is called nourishing energy, which is extracted from the purest elements of digestion and energizes organs, glands, nerves, bones and all vital tissues. The other type is protective energy, which is produced from the coarser by-products of digestion. It circulates across the surface of the body, just below the skin, protecting the entire organism from invasion by extremes of environmental energy such as heat, cold, dryness, wind, etc. When the qi extracted from food and water meets with the qi absorbed from air, they blend in the bloodstream to form the unique variety of vital energy that gives life to the human system. Since there exist no equivalent term in English for qi or prana, I shall refer to it as bionic or bioelectric energy. This combines the idea of living energy uniquely associated with the living organisms (bio-) with that of electricity (-electric) and negatively charged ions (-ionic), which comprise the essential nature of qi. Qi with a capital 'Q' refers to the sum total of all energy in the cosmos, including gravity, magnetism, solar energy, radio waves, and so forth. When it is referred of with a small 'q,' it is specifically to the bionic energy that fuels living organisms. Qi is to the living organism what electricity is to a computer. Without it, the whole complex mechanism grinds to a halt. In the case of the living organism, the polarity of Yin and Yang establishes the dynamic force field required to move qi, much as positive and negative polarity causes electric currents to move. The dynamic of Yin and Yang keep qi in constant flux, both in the external environment and within the human system. When the level of bionic energy in the body is diminished, the entire organism loses vitality and becomes vulnerable to disease, debility and premature death. The human system is strongly influenced by the various types of environmental energy, especially weather conditions, which is celestial energy. All living things stand between the positive Yang pole of heaven and the negative Yin pole of earth, and the celestial energy of weather passes through the human energy system just as electricity moves through a conductor. The presence of negative ions in the air we breathe facilitates the absorption of oxygen and elimination of carbon dioxide in the alveoli of the lungs, whereas positive ions have the reverse effect. Toxic gases, dust, chemical fumes, and so forth all take the form of positive ions when released into the atmosphere, and these big ions trap and absorb the little negative ions, leaving the air virtually devoid of vitality. A healthy body can purge itself of airborne toxins, but it can do absolutely nothing to compensate for a critical lack of qi in the air it breathes. In order to function, qi requires a strong electric field to propel it, thus the dynamic tension of Yin and Yang. The prime importance to human health and vitality are strong electric fields in the atmosphere. The absence of an electric current is always a disadvantage and always has negative effects on the vitality of the human being. The longevity of civilized man/wombman depends to a very high degree on the continual presence of a sufficiently powerful electric field. The most physiological functions - cardiac activity, respiration, digestion, metabolism, etc. are favorably influenced and stimulated by this electric field. Here we see that qi is not a static substance like blood or neurochemicals but an active force in the form of negative ions that is moved by the polar potential of electric field, i.e. by the virtue of the dynamics of Yin and Yang. Now, the next logical question is how does the body store the energy it accumulates from the atmosphere? The body stores the energy in electrolytes of all vital fluids. Electrolytes are non-metallic conductors of electricity, such as certain fluids, in which currents are carried by ions instead of electrons. The body's electrolytes accumulate and store electric charges as electric potential, then release them upon demand as active energy currents. Essence (fluids) stores potential energy (electric potential) and releases it again as active energy currents when spirit (mind) commends it. To sum up, qi is a form of bioelectric energy uniquely associated with all living things, while Yin and Yang are the two opposite poles that make qi move. The body stores qi in the electrolytes contained in vital bodily fluids (essence) and transports it through a complex network of invisible channels called meridians, which I'll thoroughly get into. The third treasure of life is spirit. Our spirit encompasses all of our mental faculties, including rational thought, intuition, attention, will, and ego. Traditional thought distinguishes four major aspects of spirit: the human soul, the animal soul, thought and awareness; intent and willpower. Western thought illustrates that the spirit is not an independent entity above and beyond the body. This division of body and mind, which runs throughout Western thought, is highly irrational and unscientific and is based more on religious dogma than on scientific fact. Contrary to this dualism, the spirit is a flowering blossom, with essence (body) serving as the roots and energy (qi) as the connecting stem. While spirit is a treasure which provides the body's guiding light, essence forms its basic building blocks, and energy is the key functional element that links them all in the great triple equation.

The Human Organism

The basic structure of all body systems, organs, and tissues is the cell. Muscles contract because their cells contract, and nerves transmit impulses when their cells are sparked into action. The cell is enclosed by a cell membrane, which is composed of phospholipids and proteins. Within the cell membrane is the cytosol, or cytoplasm. This is a fluid that distributes materials and is the center of metabolic activities. Enzymes and other proteins used by the body are produced within the cytoplasm. The cytoplasm contains an internal protein framework called the cytoskeleton, whose fibers are seen throughout the cytoplasm. Microfilaments, which are the thinnest filaments of the cytoskeleton are found in the cytoplasm of all eukaryotic cells; they provide the mechanism for contraction in muscle cells. The most important membranous organelle of the cytoplasm is the mitochondria. The enzymes for energy production are located on the inner mitochondrial membrane. High energy cells, such as muscle cells and sperm cells, contain many mitochondria. The Mitochondria is also proven to be the gene that produces nearly all the energy to keep the cell alive; transferred from the mother to the child. The center of genetic activity of the cell is the nucleus. With the exception of red blood cells and sex cells, all body cells have 46 chromosomes in the nucleus. Although there are trillions of cells in the body, there are only hundreds of different types of cells. The various types work together to form a tissue. A tissue is a collection of cells and their products organized to perform a certain function. The tissues then organize together with one another to form the body organs. There are four types of tissue that exist within the body: the epithelial, the connective, the muscle, and the neural tissues. First, epithelial tissues protect the exposed portions of the body's organs and safeguard them from abrasion and injury. They are found at the surface of body organs. They also control the passage of materials from the outside environment to the specialized body cells below, and many epithelial tissues contain sensory fibers. Many epithelial tissues are specialized to form glands. A gland is a series of cells specialized to synthesize and give off a product known as secretion. The gland may be endocrine if the secretion is distributed directly into a blood vessel, exocrine if the secretion enters a duct for delivery to a particular part of the body. Mucous, sweat, oil, and salivary glands are among the exocrine glands. Secondly, connective tissue is the most abundant and widely distributed tissue of the body. It includes bone, blood, cartilage, fat, and other tissue types that support, protect, and insulate the organs. Connective tissue is composed of cells, fibers, and ground substance. The fibers and ground substance make up the non-cellular matrix of the tissue. Muscle tissue consists of muscle cells bound together end to end to form long muscle fibers. A muscle fiber often contains several nuclei because it is composed of several cells. Lastly, is neural tissue, which contains supportive cells, called neuroglia and the impulse-transmitting cells, called nerve cells or neurons. The nerve cell is uniquely adapted to generate and transmit impulses. It contains a cell body, where the cytoplasm, nucleus, organelles, and other cell structures reside. It also contains a long extension called an axon, which ends in numerous fibers. Nerve impulses travel down the axon away from the cell body. To reach the cell body, impulses arrive by means of the treelike branches called dendrites. Dendrites receive impulses from the other nerve cells; transport them to the cell body for an appropriate interpretation.

> *The Yin and Yang energies circulate in the outer and inner parts of the body without ceasing, and in studying the state of Yin and Yang energy of the twelve meridians we are able to find the origin of illness and if there is an excess or a deficiency of energy; thus we can localize the disturbances and if we now the places where we can cause an exchange of energy between different meridians, we can reestablish the uninterrupted circulation of energy and thus the equilibrium between Yin and Yang."*
>
> —*Nei Jing, Chapter Fifty-Two*

The human body is an integral system of interrelated networks with different physiological functions. This integral system uses energy pathways to link the organs and the other human systems into a unified whole, making communication and integration between parts of the body possible. This energy flows in the pathways and extends internally to the organs and externally throughout the body, completing an interrelated system of networks. Every change that takes place in the human body produces an electrical response, just as every chemical reaction produces both electrical and chemical reactions. There are twelve important organs and bodily functions in the human organism: the lungs, large intestine, stomach, spleen, heart, small intestine, bladder, kidney, circulation-sex (function of circulation and sex), triple warmer (function of nervous system), gall bladder, and liver. Granted, there are many more than twelve separate entities making up the human body, all the other parts of the body are under the control and regulated by one or several of these twelve organs; the aforementioned are considered the primary. The internal organs are divided into two groups: the Yin, or solid organs, and the Yang, or hollow organs. Each of the Yin and Yang organs are identified with one of the five elements. The heart (Yin) and the small intestine (Yang) are associated with fire, the spleen (Yin) and stomach (Yang) with earth, the lungs (Yin) and large intestine (Yang) with water, and the liver (Yin) and gall bladder (yang) with wood. In the human body, Yin controls internal surfaces, lower regions and front parts, both on the body as a whole and on each individual organ, while Yang governs external forces, upper regions and back parts. Yin controls blood, Yang governs energy. Innate instincts belong to Yin, acquired skills to Yang. Yin descends, Yang ascends. Eating is a nourishing Yin activity, while drinking alcohol is a depleting Yang activity. In breathing, inhalation is Yin, exhalation is Yang. In the terms of seasons, cool autumn and cool winter belong to Yin, warm spring and hot summer to Yang. The Yin organs are more vital than the Yang, and dysfunctions of Yin organs cause the greatest health problems. The organs are linked by concrete functional and anatomical connections. The heart is the chief of the vital organs. The heart regulates other organs by controlling circulation of the blood. It houses the spirit and governs moods and mental clarity. The heart is paired with the small-intestine, which separates the pure from the impure by-products of digestion; it controls the ratio of liquids to solid wastes, and absorbs nutrients, which it then sends to the heart for circulation throughout the body. The liver stores and enriches blood and regulates the amount released into the blood stream for general circulation. During periods of rest, especially in cold weather, thirty to fifty percent of the body's blood supply collects in the liver and pancreas. During sleep, blood is fortified in the liver for use by the rest of the body during normal activity. The liver houses the soul. The heart and liver house our most distinctly human attributes. The liver is the body's metabolic headquarters, and therefore it is most directly responsible for a person's overall sense of well-being and vitality. The liver's Yang partner is the gallbladder. The pancreas controls production of vital enzymes needed for digestion and metabolism. This function links it directly with its paired Yang partner, the stomach. If the pancreas fails to produce sufficient enzymes, digestion in the stomach stagnates, causing food to ferment and putrefy instead of digest. The pancreas controls the human attribute of rational thought. The lungs control governing breathing and energy circulation. When breath is deficient, so is energy. The Yin lungs are associated with the Yang large intestine. The kidneys control water. Excess water and waste fluids are sent to the kidneys and converted into urine, then passed down to the bladder for excretion. Thus the bladder is functionally linked to the kidneys as their hollow Yang partner. The kidneys control the overall balance of vital fluids in the body, which in turn directly influences energy level and balance. The kidneys are the major balancers of Yin and Yang in the human system. They house the human attribute of will-power and control marrow, loins, lumbar, and sacral regions. They are also closely connected with the adrenal cortex (suprarenal glands), which straddle them and secrete cortisone, adrenaline and vital sex hormones into the blood stream. The kidneys and their related glands thus control sexual functions and potency. Now that I have given an overview of the information regarding who you are, let me explain what they don't want you to know about yourself. First let me say that it is paramount that you invest your time in acquiring information pertaining to who you are and how to utilize that information. The reason being is that when you begin to do the research and unleash the information stored in your mind you will recognize that we are truly gods and goddesses. It is amazing what I found out about myself after all the scientific research, study, etc. We are immortalist, understand that aging is a dis-ease process, and so is death. When you begin to do the research you will realize that your life has been an absurd experience. Due to the fact that we have these anatomical and neurological structures we can begin to regenerate ourselves at any particular time. Yes, you can regenerate your spinal cord, you can regenerate your brain, grow new limbs, organs, or whatever it is that you choose to regenerate. However, everything requires first of all, an acknowledgement that this is possible and then secondly to find out precisely what you need to do to actualize the experience. So anytime you begin to collect thoughts in your mind that things are imposssible; that experience are not experiences for you for reasons of primarily inadequacy, then do overstand that obviously by universal law those things cannot be in your realm of reality. When you begin to evaluate your belief system and the foundation upon which you define yourself; if you are holding on to belief systems that just by the thought of them create discomfort, poverty, thoughts of what you can't afford, or what you can't do, why you have to continue to be on jobs that create discomfort for you everyday, etc. Overstand that these are all experiences only based on your belief system and what it is that you are allowing yourself to entertain. The Universe, and especially in this dimension, which is known as a holographic dimension; material world, very much so a mirroring dimension. Whatever you believe creates your reality. So therefore when you have belief systems that immediately create discomfort, you have to overstand that by choosing to continue to think that thought that your create your own discomfort. It is only what we call a relative reality; it is not an absolute reality. When I say absolute reality, I'm talking about the laws that permeate all dimensions. Any dimension you find yourself in, you have to obey those absolute laws, but on a relative sense; relative reality has really been cloaked and has been perpetrated by social and cultural norms; which are chemical realities that are actually written upon in your genetic structure. Now, many belief systems are 300,000 years old. It has been discovered that Lucy's genes are actually 500,000 years old. Some interesting things will come out about that eventually; like how she really didnt even originate from this planet. How awesome is that? When you look at the genetic count of the Aboriginies, which is actually forty-two plus two chromosomes. You have to overstand that we all don't have the same chromosome makeup. There's forty-two plus two chromosomes, forty-four plus four chromosomes, and forty-six plus two chromosomes. So depending on the chromosome number that you have and also the number of chromosomes that you are actually able to activate depends and will determine your reality and the capacity you have to master this dimension. When you begin to study you will realize that there are actually seven races on this planet that are divided based on their chromosome number. The aboriginies are the chromosome that the modern scientist have identified, which has a forty-two plus two chromosome makeup; containing at least 300,000 years of information in their chromosomes. Unfortunately they have been one of the groups of people that have been researched on most intensely for that reason. Contrary to what you've been taught, we have twelve sets of double helix's not two. Most people don't even activate ¼ of one single double helix, which is why they are experiencing the kind of life they have. Limitation, lack of any type, or discomfort is not your destiny. Do not ever allow a person to indoctrinate you into thinking that. If you are experienceing it, its primarily due to the fact that you have a relative reality not aligned with facts and truth. Truly, we are all potential gods and goddesses. That is a state of existence that you have to personally earn, but the potential was already been given to you; not until you are ready to do the personal work and begin to overstand who you are can you begin to activate the seven levels of brain function. You have seven circuits in the brain that you have to activate; you need to learn about your neuroanatomy and truly learn how to master each circuit.

The Five Elements

The Five Elements theory posits wood, fire, earth, metal, and water as the basic elements of the material world. These elements are in constant movement and change. The complex connections between material objects are explained through the relationship of interdependence and mutual restraint that governs the five elements. Ancient physicians used the five elements theory to study extensively the connections between the physiology and pathology of the organs and tissues and the natural environment. By adopting the methodology of comparing similarity to expose phenomenon, the ancients attributed different phenomena to the categories of the five elements. On the basis of the phenomena's different characteristics, functions, and forms, the complex links between physiology and pathology as well as the correlation between the human body and the natural environment were explained. Now in order to master the body, one must gain control of the inner and outer Universe. Overstand the dynamics of the Universe, Planet Earth, and the human body in respect to their relationship to the five elements -

water, fire, wood, metal and earth. There is an interrelatedness of all things through the use of those close relationships. Each of the five elements has this type of mutual generating relationship with the other. Mutual generation means multiplication and promotion, while mutual subjugation means mutual restriction and restraint. In this way generation is circular and endless. In the mutual generating relation of the five elements, each of the elements has the property of being generate and generating. The one which generates is the mother, the one which is generated is the son; this is known as the mother-son relationship. The five elements produce each other, wood creates fire since fire results from rubbing two pieces of wood together and wood burns easily. In leaving ashes which become part of soil, fire creates earth. Metallic ores are found in the earth, thus earth creates metal. Metal creates water because when metal is heated it becomes liquid. Lastly, water creates wood by nourishing the growth of plants. Also between each of the five elements, there exist the close relationships of mutual generation, mutual subjugation, extreme subjugation, and counter subjugation. According to the order of mutual subjugation, however, wood subjugates earth, metal subjugates wood, etc. Each of the five elements also shares this subjugation relationship with the other. This relationship has the properties of being subjugated and of subjugating. For example, wood weakens earth by removing nutrients from the soil. Earth limits water by natural bodies of water, such as lakes and rivers, and dams. Water extinguishes fire. Fire conquers metal by melting it. Metal, in the form of axes and knifes, can cut down trees and carve wood. Conversely, an element respects or fears the element which could destroy it. Wood fears metal, metal fire, fire water, water earth, and earth fears wood. Therefore, the mutual subjugating relationship among the five elements is also known as the relationship of being superior to and being inferior to another element. Mutual generation and mutual subjugation are two aspects which cannot be separated. If there is no generation, then there is no birth and growth. If there is no subjugation, then there is no change and development for maintaining normal harmonious relations. If there is no restriction, then endless growth and development will become harmful. Thus the movement and change of all things exists through their mutual generating and subjugating relationships. These relationships are the basis of the never ending circulation of natural elements. For example, water has the properties of soaking and descending (water flows downward). Fire heats and moves upward (flames rise into the air). Wood allows its form to be shaped into curved or straight pieces. Metal can be melted, molded, and then hardened. Earth's properties include the provision of nourishment through sowing and reaping. The same pairs of elements are related to each other by the principle of mutual closeness. Each element is considered attracted to its source. Thus wood is close to water, water to metal, metal to earth, earth to fire, and fire to wood. These similarities and differences among the principles have been analyzed for thousands of years in terms of yin and yang. Creation and closeness, both constructive principles, are Yang; whereas, destruction and fearfulness, their opposites, are viewed as Yin. In addition to representing forces in the natural world, the five elements provide guiding principles for physiology, pathology, diagnosis, and therapy in traditional medicine. The elements, together with Yin and Yang, will determine the state of balance and equilibrium within the body. Extreme subjugation and counter subjugation are the pathological conditions of the normal mutual generation and subjugation relationships. Extreme subjugation denotes that the subjugation of one of the five elements to another surpasses the normal level. For example, if there is hyperactivity of the wood element, it will subjugate the earth element. The latter element is made weak and insufficient. Counter subjugation means that one of the five elements subjugates the other opposite to the normal mutual subjugation order. For example, when metal is weak and insufficient, it leads to the hyperactivity of wood. When the qi of one of the five elements is excessive, it will subjugate its subjugated element (such as wood subjugating earth) and counter subjugate the subjugating element (such as wood counter subjugating metal). Moreover, the Five Elements theory recognizes a correlation between those things which are related to a particular element; each element has its own repertory of relationships among the objects that compose the physical world. The theory of Five Elements is therefore the theoretical basis of the unique bond between man/wombman and nature.

The Twelve Meridians

There are thirty-two known meridians of which twenty-four belong to the organs and bowels while the other eight govern other functions of the body. Each organ and bowel is governed by two meridians: one flows from the left and one flows from the right. The pressure points are the breathing points for the meridians. Think of the meridian as a train route. The pressure point is like a train station and the commuters are the energy which flows along the meridian. The energy gets on and off at the train stations. The twelve meridians form twelve lines, which are associated with the twelve primary organs. Each meridian is paired, one part being on each side of the body. There are three upper Yin meridians, going from the chest, down the inside of the arm to the tips of the fingers. They travel from inside to outside the meridians of the heart, circulation-sex, and lung. There are three upper Yang meridians, going from the tips of the fingers, up the external surface of the arm to the face. They travel from inside to the outside of the meridians of the small intestine, triple- warmer, and large intestine. There are three lower

Yang meridians going from the head, down the body and the external surface of the legs to the toes. They travel from the inside to the outside of the meridians of the stomach, gall bladder, and bladder. There are three lower Yin meridians going from the toes, up the inside of the legs, over the abdomen ending on the front of the chest; near the upper Yin meridian, traveling from the inside to the outside of the meridians of the spleen, liver, and kidney. The arrangement of the meridians follows the basic laws of embryology and comparative anatomy, meaning that the evolution of the animal kingdom repeats, in a certain way, the evolution of the embryo; similar to the arrangement of the various layers of cells in the earliest stages of embryonic development. Overstanding the nature, the connection, and the pathway of this energy will enable one to perceive the connection of three bodies; physical, soul, and spirit within each individual, ultimately fusing them into an immortal body. There are two liver meridians. They belong to the wood element and are Yin property. Each liver meridian has fourteen pressure points. The liver meridians control the functions of sex organs, digestion, urination, chest and the inside of the legs. They also govern the liver. The gall bladder meridians are in charge of the gall bladder. Each has forty-four pressure points. Each commences from the corner of the eye running right down from the neck to the chest, down the side of the body, the outside of the leg to the little toe. The gall bladder meridians are responsible for the head, the eye, the ear, the nose, the mouth, the throat, fever, chest, outside leg and lower leg. They are the family of wood and Yang property. The heart meridians each have nine pressure points. They are the element of fire and are Yin property. They are a means of controlling heart disease and insomnia. The small intestine meridians also belong to the element of fire and are Yang property. Each has nineteen pressure points. The circulation-sex meridians also belong to the element of fire, and are Yin property. Each meridian has nine pressure points and deals with heart trouble and loss of memory. The triple warmer meridians are the element of fire, and are Yang property. Each meridian has twenty-three pressure points. The spleen meridians belong to the element of earth and are Yin property. Each has twenty pressure points. The stomach meridians belong to the element of earth, and are Yang property. Each meridian has forty-five pressure points. The lung meridians belong to the element of metal and are Yin property. Each meridian has eleven pressure points. The large intestine meridians belong to the element metal, and are Yang property. Each meridian has twenty pressure points. The kidney meridians belong to the element of water, and are Yin property. Each meridian has twenty-seven pressure points. The bladder meridians belong to the element of water, and are Yang property. Each meridian has sixty-seven pressure points. The therapeutical aspect of acupuncture; with stimulating, usually with a needle, of specifically indicated points, with the object of tonifying or sedating, i.e. increasing or reducing the quality of energy in the particular organ or meridian which is affected. It has already been pointed out that the meridians represent the course along which flows the energy associated with the various organs and bodily functions, so they can form, as it were, a kind of chart of the body's energy pattern. Nearly all the points, known as acupuncture points are located along one or the other meridians. In all, there are about a thousand acupuncture points. I just discussed the primary of them. In addition to these, there are a number of traditionally accepted points located outside the meridian and some newly discovered points making a total of about seven hundred eighty-five points. These points are well known and described in the literature of acupuncture. There are in addition well over two hundred minor points of secondary importance so that one thousand points would be a rough estimate of the total figure. The points are capable of powerful reaction and should not be regarded lightly, but through study application becomes autonomous.

Water

Water is vital to everything about you. Your cells sit in water; they are made up of or composed of water. Your brain even sits in water. Water is H2O; a hydrogen atom and two oxygen atoms chemically bonded together to form the molecule water. That chemical bonding, that exchange of electrons from atom to atom produces what we call electricity, or the moving of electrons from molecule to molecule produces electricity. Electricity is the flow of electrons. In order for your body to properly function, no matter what type of system (they say circulatory, but all of them are circulatory systems), they all go in a cycle or a circle; the digestive system, respiratory system, nervous system, etc. All of these systems run off of the flow of ions, or electrons in the body. For example, your nervous system; your nerves have impulses. You think it, and there is an impulse or an electrical flow of electricity that leaves your brain through the nerve or neurons, and they stimulate your muscles to move; that is electricity. Your body thrives off of electricity. When you cut the flow of electricity or ether off, because electricity is a form of ether; when you cut that flow of electricity from your body; reduce it, or if you hinder it, dis-ease manifest in your body. That electricity is the medium to the etheric realm for your body. As long as the environment is right for electricity to flow within your body, you are fine. When you cut that off, you're not. Imagine that fatigue, energy depletion, depression, eczema, rheumatism, high and low blood pressure, high cholesterol, and premature aging could all be caused by a chronic lack of water in the body. The food that most people eat is too rich, too concentrated, and too salty, and the use of dehydrating substances such as alcohol and tobacco is very widespread. Water is vital because it gives you free ions or free electrons for your body to use in the different processes it performs within the body. The proper water that has electron flow within it helps to stimulate and maintain electricity in your body. When water has a high pH it has more electron flow in it, or electrolytes. Hence, it is more electric; which means that the water is more conducive for electronic flow or electron flow we call electricity. Although the body is constructed of both liquid and solid materials, fluids are present in much greater quantity than solids. Physiology teaches that water is the most important constituent of the body, accounting for seventy percent of the human body's composition. The fluids of the body are separated and allocated to different compartments throughout the body. The blood is the fluid closest to the body's surface. The blood is the first to receive substances taken in by the body from the outside, such as oxygen brought in by the respiratory tract and nutritive material passed through the mucous membranes of the digestive tract. The blood represents five percent of the body's weight, yet it circulates only within the arteries, veins, and capillaries of the vascular network. Extra cellular fluids and lymph receive oxygen; fluid form, and nutritive substances carried by the bloodstream, and then it transports them to the cells, where it is utilized. The extra cellular fluids also receive the waste products and residues produced by the cells and transport them to the bloodstream, which in turn takes them to the excretory organs (liver, kidneys, etc.), so they can be filtered and eliminated. The lymph fluid removes a portion of the toxins it has absorbed from the cells and carries it up to the bloodstream. The lymphatic vessels in which lymph circulates, spill into the blood at the same level of the subclavian artery. From this point, the toxins are directed toward the excretory organs. Together, extra cellular fluids and lymph represent fifteen percent of body weight, a weight three times that of the blood. The intracellular fluid is composed of all the liquids located within the cells. The fluid represents half the weight of the entire body. Oxygen and nutrients are carried by the extra cellular fluid that penetrates into the intracellular by traveling through the cellular membranes. The lungs and the heart, for example, consist of seventy percent water; the muscles are seventy-five percent water; the liver is seventy-five percent water; and the spleen is seventy-seven percent water. The brain is the organ with the highest fluid content, eighty-three percent. With the exception of a few organs or body parts (the skin, the nails), the concentration of solid substances is seventy-eight percent for the skeleton, cells acquire their solidity from the water that fills them. The water that fills the cells exerts pressure on the cellular envelope, which gives the cells their shape and solidity. Water travels through the body in three stages. The first stage is through intake, then absorption by the tissue and cells, and lastly; elimination. Every day, we ingest around 2.5 liters of liquid by mouth. This liquid is either in free form or bound within the tissues of the solid foods we eat, such as juice contained in fruits and vegetables. A person's water intake depends on the foods in their basic diet. Certain foods don't possess a sufficient amount of water essential for maintaining a balanced diet; grain, for example. Certain foods that are dead don't have any hydrogen or any ions in them. Not any, but they are low in energy. That energy is atoms, or fluid, or movement of atoms of atomic structure. Furthermore, given the mentioned explanation of what water is, what role it plays within our body, and the cycle it travels; water is used in every cell of your body. It is the main component of the fluids in our body, which carries nutrients, oxygen, and waste from our cells and organs. The majority of the parts in the body need water-based fluids to survive. Water is part of your body's temperature regulating system; it cushions joints, protects tissues, and organs like the spinal cord. Water aids in digestion, absorption of food, and aiding in the removal of waste from the body.

Nutrition

Now, in the past four hundred and fifty years, our anthropologist, our historians, our revolutionaries they never talk about the very important thing that is supposed to be addressed; our diet. Everyone is talking about spirituality, and things. What is that? What is spirituality? How do we use that to prepare us to enhance our life? How do you use it? Society shows you all sorts of things to believe in cause we live in the world of belief. No, I don't live in the world of belief; either I know or I don't know. So now, in the last 450 years western medicine has said when you are sick you have a germ, a virus, or bacteria; which is the cause of your manifestation of disease. If they have identified the cause of disease how come they are not curing anything? Now, there is only one disease; mucus. Anything that you put in your mouth that is acid will eat your body up. Like for instance, wheat, rice, beans, cow, hogs, and chicken. All that stuff is garbage. The Dutch made the carrot. The Portuguese made the casaba, and the British made the bean; now we are a slave to all of this. You will never find a gorilla in Alaska, right? Nor a coconut in Canada, why? The reason being is because the Universe is based on a cosmic arrangement. On Earth, this arrangement has been violated. You and I are victims. We've been eating the wrong food because we didn't know what to eat. We have been drugged with glucose. Glucose is a chemical that will change your thinking patterns. It intercepts and confounds the hypothalamus in your brain. That's what sugar does; starch also. Something went wrong. We were unaware that the stuff they gave us to eat was under minding our thinking. We are a beautiful people; we are a loving people and smart people. I'm not smarter than you. How can I be smarter than you when I carry the same amount of gene cells in my brain as you can? I'm not smarter, I just have a message. Some of you have a message that I don't have. We need all of these to enhance life and to protect our own existence. So now, we need herbs and food that are not made by man. What are the herbs made by man; aloe vera, peppermint, garlic, and nutmeg, etc. Just a few example. Nutmeg has arsenic in it; it was made, therefore unnatural. The mistake we are making is that we haven't been addressing genetics. What is genetically consistent with a group is not so with another. Like I said, a gorilla doesn't eat a polar bears food, so why is it that we eat everybody's food? When they took some of us from Africa, did they bring their food with them? No. Well then what have they been feeding us? The aboriginal has always clanged to the laws of life, not philosophy. Philosophy has no place in life. Let me clarify, let's start with protein. This word protein came from your parents, from your brother, from your gene and from the European culture. Protein is not a Moorish word. What is protein? Protein is 1 of the 9 amino acids; the building blocks of life. Let me remind you that the human body has no use for acidity. The body has no use for acids; where do they fit? The body has 102 minerals. When one of those minerals is depleted your body becomes sick. If iron becomes depleted, you are anemic. If calcium becomes depleted, you have remnants of bone structure. If you cannot think you are lacking in carbon and copper. So where does protein fit? Furthermore, is it electrical? I mean, the body is electrical, it moves. I cannot move unless my body is electrical. There could be no motion if that motion isn't electrical. So, if you're going to eat protein it better be electrical because the body only assimilates a substance through chemical affinity. So, once again is it electrical? We've been bamboozled as Malcolm said, but we didn't know to what extent. Enough is enough. We have been fed the wrong foods. We are not supposed to eat meat, it will hurt you. If I have not traveled the journey how would I be able to take you? I had to travel through it. It's not mystery, it's not spiritual; it is natural. When we were living in Africa and North Amexem, before the European came we didn't have such words as spirituality in our vocabulary; it wasn't needed. We didn't have any religion because that wasn't needed. We were living in accordance with life and the Creator because that's all we knew. Just like the animals of today, we need to live by the laws of life. When you start to think about eating something you need to ask yourself; is it consistent with my biology and is it electric? We have been disconnected with life; or there has been a severance of some sort. It's the food that's killings us; it makes you sick, then you go to the doctor and trust that they will cure you; how crazy is that? It really is insanity. Now when it comes to healing there are several expressions or ideologies. You have the allopathic, which is based off of carcinogens; homeopathic, which is based off biochemistry, and they represent European ideologies. There is also macrobiotic, which is based off of rice; a hybrid made from starch and cyanide that kills mucus membrane, and it represents the Japanese ideology. Next is the Chinese ideology of healing which is based off of ninety percent cultivated herbs; hybrids, or artificial plants that are acid base. The Ayurvedic ideology is based off of ghee and garlic. Now ghee is lactose which cannot assimilate in the body and garlic is an oxide that burns cells and that's why you feel that burning feeling when you ingest it. I'm aware that most have used garlic for years and its known to lower pressure, but yet it destroys cell membrane. Now, if you've noticed there is one philosophy missing, which is the cornerstone that the builders rejected; but yet it is the one that completes the building. When you begin to investigate these philosophies; before mentioned, you will clearly see that each one of them presents us with a substance that instead of complimenting us they do just the opposite; they offend our biology. Now to reiterate, carcinogens erode cells because they are acid base. They have a tendency to disrupt the hormonal flow of the body; causing stress and loss of energy. Biochemistry, which is homeopathy, analyzes the body according to the minerals that are missing or deficient, and that determines the disease. For example, the brain correlates to copper, the blood to iron, and the bones to calcium, etc. Now a lack of iron indicates anemia, sickle cell anemia, hypoglycemia, leukemia; and lack of calcium indicates osteoporosis, etc. If your iron is lacking you cannot think clearly because iron is the mineral that conveys oxygen to the brain. Instead of offering iron that is consistent with human biology; iron that has carbon, hydrogen, and oxygen, they decided to use a substance, or supplement to replace the minerals that were deficient; an oxide, which is a rock, or rust, which consists of hydrated iron oxides $\{Fe_2O_3 \cdot nH_2O\}$ and iron oxide-hydroxide $\{FeO(OH), Fe(OH)_3\}$. Furthermore, they use it to make steel; the body cannot assimilate a rock, overstand? So I say that to say truth is the answer to nutrition, but we have so many differing expressions of it; Hinduism, Buddhism, Judaism, Christism, Muhammadism, etc. One thing you'll notice is that each consumes substances that the creator didn't make. If one could easily conclude, that there is one Creator; there should be only one school of thought representing that one Creator, but instead we have many expressions. By you indulging in things that aren't part of creation, it divided us from ourselves. When someone consumes meat, rice, beans, cheese, butter, and eggs; by the time they reach thirty, there is already the manifestation of disease. The mucus membrane has been compromised by the acidic food you've been ingesting. The brain suffers first and thought patterns change.

Further, there is stress one is presented with and all because you're out of balance. It's like putting gasoline in a diesel car; or feeding a polar bear grass and a gorilla meat. For instance, the cows in England were given meat and flesh with the concentrate they were eating. What happened to those cows? Mad cow disease, right? This is what you're doing; eating cows and expecting to be sane. So, could that be the reason for all the division and confusion? The Creator didn't make a cow, it's a hybrid animal. They are not a product of life; but because you have disconnected it is difficult for you to see that it is even possible that an animal could have been made; but of course, through grafting, just like oranges and grapefruits. Tofu is made in England, which George Washington Carver used to make plastic. The soybean is a starch and created by a Benedictine priest, Menden I think his name is? What nutritional value does plastic have? Hopefully, with this little bit of information you're prepared to dive a bit deeper and invest in your health. For us today, the most important thing is nutrition. What is the real diet of the Gods and Goddesses? You are supposed to eat mentally, and in doing so it becomes chemicals in your body; you think and produce chemicals. You eat food, or what you're calling food, and it affects your ability to think;

either on an extremely low frequency, waves, or on a resonating high frequency. Every emotional behavior has been identified electrically, scientifically by our people in the past; and now presently by Europeans. I cannot stress that the most important thing is what you put in your mind. There is a particular science to manifesting the genome of the aboriginal. If you violate those laws, you start hybridizing your consciousness and eventually you produce a hybrid being. Our ancestors made him and her. They're geometric verification of our disobedience and transgressions against absolute law. His real name is the devil. I am not name calling, nor am I a racist; Malachi said, "No one wins the race in racism", I'm a scientist. Devil means grafted or hybrid, a live germ grafted from original. A grapefruit is a devil. An orange, is a devil, and a watermelon is a devil. These are just some of the foods that were grafted, or made in a laboratory. You need to overstand what is operating on the quantum and cellular plane within your body, and how it operates cosmologically. The cells in your body operate on the principle of absolute law. It is the feminine and masculine principle. They operate in symbiotic relationship; that is the foundation of all existence, we call it love. This is the foundation of every operation of the universe.

The body operates off these principles as well; a left, a right, a up, a down, etc. It is called duality, yet it is oneness; polarity. The body works off of electricity and magnetism; you have two systems, one masculine and one feminine. The feminine is called the endocrine system which deals with hormones, and a masculine called the nervous system which creates neurons. These realities are produced in your brain and communicate to every cell of your body to tell every cell what to do. They are produced in the divine triangle, which is called the pineal gland, the hypothalamus gland, and the pituitary gland. Each of the glands secretes neurons and hormones that communicate to every cell of your body. The chemicals are produced by electricity and magnetism; light frequencies, ether, air, and food; the four elements of creation, air, fire, water, and earth. The pineal gland is a GPS, it picks up wave formations and turns them into chemicals; serotonin during the day which is your masculine principle, and melatonin at night, which is the feminine principle. Now, the pineal gland is the gland that makes law in the body. You cannot live as an aboriginal being without it. The hypothalamus gland is the gland that executes law; it connects the neuronal system, nervous system, and the hormone system. The hormones and neurons are communicated to the pituitary gland, which is the master gland of congress in the body that sends hormones to every cell in the body. These are your three branches of government; congress, which makes law, the executive branch, which executes, and the judicial branch, which interprets the law. What is happening is that consciousness is not ruling from the kingdom in your body. There are foreign invaders in our body; do you know what they are called, viruses. Your food is made of viruses. I know you think I'm lying, but I love you and that's why I'm sharing this with you. I can quote the European, I'm sure most would like that. We as a people; if you study your history can find the point at which our ancestors began to violate the law. It began when they left the proper relationship between man and wombman, and when they left that relationship in the body they created an overabundance of acid. For those that don't know, the body is to remain alkaline, which means electrical. When the body is acid it cannot detect foreigners in its presence; foreigners being anything unnatural. Your body is made of hydrogen ions; if you have six hydrogen ions we call it carbon, if you have eight, we call it oxygen; seven we call in nitrogen, and if you have nineteen, its potassium, etc. They are all positive hydrogen ions which means photons or materials; crystals, that's what your body is made of, minerals and elements. Negative hydrogen ions are electrons, which are energy fields and minerals or elements making up who you are, thus the feminine and masculine principle. In order to stay alkaline you must have more electricity in your body than you have acids. Overstand? I know some people think we're a matriarchal culture, and others a patriarchal culture, but yet we're neither. You see, their objective is to place you into polar shifts to fight and not overstand that man and wombman are one being; you cannot have one without the other. The Eurocentric, devil minded, grafted concept is to compare the two and juggle between matriarchal and patriarchal; this is how our people became confused. In Sumeria, we started exploring with production methods of agriculture, or farming. Now, what is the etymology of the word farm? It comes from the Greek word paramecia, which comes from a word in the Sumerian language, cuneiform and the word meaning sorcery. At one time we were a people who knew and overstood that in order to maintain the health of our genome the earth would produce what we needed; the so-called garden of Eden, where food is produced in nature, naturally. When our ancestors started farming in Sumeria, Iran, Saudi Arabia;

they created acidity in the soil, altering the consciousness of the plants. They produced foods using yeast leading to fermentation, and when this occurs in the body, cells rot. This is what is happening to us, we rot away to cancer, HIV, diabetes, cerebral palsy, etc., we as a people are deteriorating because our consciousness as a nation of people is not on a level it's needed to be. Now, our ancestors produced yeast, which is a hybrid hormone in a plant, and from that changed the genetic structure of plants; the plants began to die and presently the land known as the fertile crescent of civilization is a desert. It use to be the most beautiful and plush land on earth. Why is it a desert? Why is the Sahara there when it use to be filled with agriculture, or produce? The reason is because our ancestors started living a lifestyle that's off the frequency of aboriginal people. To go back, the word acid comes from the ancient origin Ast, or Aset; which is the throne. The throne is the materialization of the idea of God; the materialization is the physical, gross, dense part of creation; it is the acid, but it is always carried a symbiotic relationship with alkalinity, which is electricity. Therefore, you have what we call good health. When they started operating in matriarchal rule, our ancestors became vulnerable, because a woman is not by nature a military master. I know most sisters are thinking brother that's some Eurocentric patriarchal philosophy, but it's not, it is the truth. You have another aspect of operation that you deal on to maintain the order of the first government which is family. Now yeast is the real code name for virus; another name is enzyme. It's funny listening to these ignorant overseers that tell people, "hey brother, we need enzymes and protein" It's obvious they don't know the difference between an aboriginal body and a hybrid body; listening to these European doctors, telling you that you need enzymes and all this garbage. Why do you trust them? I'm your brother. It's ok, I know, you just don't want to study. So inside the cell, when the cells get acidic and there is no communication coming from the kingdom, what happens? The yeast inhibits the Mitochondrial DNA; once it gets in to the Mitochondrial DNA, which is the engine of the cell, it begins to take on instructions from the foreigner. The feminine aspect of the cell begins to take on instructions and becomes seeded by a foreigner, sounds familiar, huh? The masculine part of the cell, the DNA, which communicates to the Mitochondrial DNA by RNA is now shut down and subject to the instructions of a foreigner; an extremely low frequency biological entity and its own feminine principle, which can now have the power to subdued it into whatever particular entities decree. Sound familiar? So my people you have lost your nationality in substance and form, and we as a people are sick. We did it to ourselves but it can be rectified and it must be done now if things are going to change. Eating improper food allows your cells become a so-called dysfunctional family. The children that come out of this are cancerous cells, sounds familiar? Hopefully you overstand the basics of nutrition which also implements Moorish science. Now if you eat meat, it's in your body. Right now, if you eat cheese, it's in your body; soy products, that's in your body, and any other acids, etc. In doing so, you have an inferior consciousness operating you, and you are not mentally alive. We as a collective nation are not mentally alive; it's basically what they call the mentally dead. Let me explain something to you about your so-called friend the European scientist. He goes into your cells microscopically; they're using yeast which is a fungus, a heterotroph which means it doesn't need light to grow, it doesn't like to grow in light, and it's not phototropic like us; with is a deep carbon hue. Furthermore, in order for those scientists to do the things that people think are so amazing with the human genome project and study the molecular nature of the cell; they had to us foreign invaders (yeast) to go into cells to actually get a view of what's going on in the cell. They learned how to place hybrid cells inside of food, which means when you get it, you're eating the hybrid DNA and now your DNA structure is altering and changing, and is no longer original and you don't even have the ability to think on a level to solve our problems as a nation of people. The European scientist uses dyes and fermentation (yeast) to swell up DNA so they can actually study it, and in doing that process they've never seen a living gene because their consciousness and their technologies won't allow it; so they can't even study life. Yeah, they've been out here writing all these books, and the ignorant are out here reading them, giving it validity, but yet you question me and us; your brothers and sisters. It's ok, I know it's the glucose that's got your mind in a state of confusion, it's alright; I forgive you for your lack of compassion. Now, back to these European scientists; we're dealing with mad scientist. They are trying to figure out how to take genetically modified DNA, place it in a living tissue and activate it on an extremely low frequency, but still have it alive; sounds like Frankenstein, huh. They're trying to redo the process that created them. They need the mitochondrial DNA of the cell to even be in a state to receive the foreigner, i.e. acid. This is why they come after our woman. Once they place the artificial virus DNA in a cell, it behaves as the natural one did. The virus normally multiplies in ecoli; did you know spinach was made of ecoli? I didn't either, until recently. I know you've heard of ecoli outbreaks in spinach, haven't you? Spinach is a hybrid food; you should never even touch it. Please don't get upset, I'm telling you the truth. This is the level of science and mathematics the European scientist is willing to go at to attack us; what are you going to do? Are you willing to study? I'm thankful that you've even made it this far. Gratitude. Now, eighty percent of the food is made of viral products, called recombinant DNA. Now according to nutrition we as a people are in obstruction of Moorish science because we do not live according to the proper symbiotic relationship between the masculine and the feminine principles, and therefore we let foreigners into our family and it disrupts and destroys the family. So the decision lies between being willing to become one, like we really are, or will we as a people, as a nation except this altered form of consciousness that we're adversarial to one another and continuing to fight amongst ourselves. Do you really want to make a change, or are all form and no substance? You have to be willing to really change yourselves, and teach your children to live by mathematics, which is to be right and exact.

Yoga and Meditation

These days, all of a sudden it seems yoga is a popular mode for exercise, a hot trend of physical postures often performed by these so-called glamorous celebrities that promise to lead you to greater fitness and ravishing beauty. But be assured, however, that there is a lot more to yoga than some passing fad. The roots of yoga can be traced back over 5,000 years. Some say early writings of yoga were transcribed on palm leaves, but due to their fragile nature, little evidence of the beginnings of yoga remains. It is also said that the beginnings of yoga were developed by the Indus-Sarasvati civilization in Northern India. Yoga was mentioned in the Vedas, which is one of the oldest sacred texts. Over the years, Vedic priests refined and documented their practices of yoga in texts such as the Bhagavad-Gita, but it wasn't until Patanjali's Yoga Sutras that yoga was systematically presented. In a confluence of Eastern mysticism and Western science, doctors are embracing meditation not because they think it's hip or cool but because scientific studies are beginning to show that it works. Despite its amazing growth in popularity in western society, many serious practitioners of the ancient art of yoga and meditation see it as nothing more than a series of powerful physical exercises designed to give one a perfect body. First and foremost, yoga is a systematic process of spiritual unfolding. The path of yoga teaches us how to integrate and heal our personal existence, as well as harmonize our individual consciousness with creation. Yoga philosophy should beat the very heart of any good yoga practitioner. Meditation (Dhyana in Sanskrit) is a specific term in yoga philosophy. It is different from the term meditation used by most people. Yoga meditation is the penultimate stage in a progression of stages towards the final state of yoga, namely Samadhi or yogic trance. For meditation to happen, Kundalini needs to have risen along the spine and reached the Sahasrara chakra at the crown of the head. Then you sit and practice keeping the four lobes of the brain peaceful. In other words, meditation is similar to ordinary meditation, but not just the thinking faculty is at rest, and it is only to be attempted after Kundalini has risen. There are two approaches to having Kundalini rise: the traditional yoga path starting with Asanas (postures) and progressing to Pranayama, Pratyahara and Dharana. The other approach is not just to practice the eight-fold path but also to perform Ishvara Pranidhana, surrender to the creator. The human being is part of nature and nature's driving energy is Kundalini. Yoga practice and sex in combination change Kundalini from being a potential energy from which all other energies in the body draw their power, to being an active energy to be utilized in your quest for vision of and contact with God in the form of divine energy in the Sahasrara, or crown chakra. The mind has been divided into four parts as per actions in philosophical and yoga texts as follows: First, there is the Ego: the feeling of body being the Self and that all actions are connected to this Self. Human beings do actions based on this blind notion. Sensitiveness of the human being comes under this part of the mind. Control of this part and the eventual removal of this notion through meditation is the ultimate objective. Secondly, there is the Intellect: the capacity of the mind to differentiate the good and bad and make decisions. The will power to take actions comes under this. With this part only, meditation can be achieved. Thirdly, there is Mind: this is the energy part that controls all the body movements automatically. The five intellectual senses and five working senses come under this. Control of this part is the first objective of meditation. Lastly, there is the Soul: this is the part where all attitudes and impressions which one brings along with him at the time of birth. This is also the storehouse of past lives and present life memory. One's action in this life depends upon the impressions encrypted in this part; these impressions can only be erased through meditation. Now, when you begin meditation, the mind wanders here and there and it never stays on one object. Consequently, one cannot get the desired concentration because when a child is born it brings along with it all the impressions of its previous births both in character and action. It also brings along the ingrained good and bad karmas. From the time of birth until it is able to talk, a child is considered to be innocent. Slowly the parents inculcate into the mind of the child what is good and what is bad as per their customs. Not only do the social circumstances in which the parents live make the child understand and modify its mind as needs require, but it also creates additional impressions in the mind. By the time the child becomes an adolescent, it slowly begins to desire; ambitions come into the picture, which force the mind to adopt new attitudes. It is further contaminated when the child matures sexually and furthermore, as sexual desires dominate the adolescent person. Attraction to the opposite sex comes automatically and this results in various actions from the youth, steps which may or may not be good. Disappointments start from the time the child goes to school and builds up gradually in the mind with impressions. Sorrow begins to show and this causes bodily afflictions, if not immediately but at a later stage; then comes employment and the strains that accompany employment. Marriage and children increases responsibilities. Karma will make you to suffer as you are destined. Some may find less sorrow due to their good karma but the majority suffer due to bad karma. Thus, in life, one finds many ups and downs. One enjoys the pleasures and also suffers the pains. Mind's natural tendency is to seek pleasure. It cannot withstand disappointments of its desires and when one desire is not fulfilled it feels sorrow resulting in pain. Thus, gradually, one finds themselves burdened with many pains which remain ingrained in the mind. Along with this, all the actions and all the scenes, such as what one has heard, one has seen, one has touched, one has smelled and tasted also remain in the mind as impressions. Some may remain permanently in the mind just like in a hard disc in a computer. Others will remain as temporary files in the computer and get automatically erased. When a person meditates;

impressions remaining in the mind start appearing before the person. Those who teach meditation will say, "Remain detached and watch all that comes before you." This is more easily said than done. It will take practice to keep the ego away from all these scenes and watch them dispassionately. You cannot derive any benefit as long as you watch these with attachment of ego. Only when watched with detachment do these impressions unwind themselves. When sitting for meditation, it is advisable to test the flow of breath. After finding the flow, first see that the entire breath is exhaled and close the nostril in which there is no flow. For closing the nostril, it is recommended to use the thumb for the right nostril and the little and ring finger together for the left nostril. Take as deep a breath as possible without straining the flowing nostril while closing the other one. Then close the nostril through which you have taken the breath and release the finger from the other nostril and exhale. Through the exhaled nostril, take a deep breath inhaling and close that nostril and open the other nostril and exhale and again inhale and repeat the same way. One inhaling through one nostril exhaling through other nostril and inhaling and then exhaling through the first nostril is one full Naadisuddhi Pranayama. One can do this practice either with closed eyes or with open eyes. This is very essential to purify the channels and ultimate results of meditation can come only after purification of the Naadis(also known as energy channels). Once this part is over, gradually close your eyes. Try to concentrate on the centre of the forehead. There are various types of meditation that are being marketed in the western world by various ashrams and their disciples, who have branched out on their own. There are genuine teachers that help humanity with free discourses and these need not be doubted. Meditation is not a simple thing. Everybody is exhorted to sit in a straight pose and close eyes and concentrate on a particular section or a symbol. It is not as easy to practice. In meditation the ultimate objective is the state of meditation, which means one-pointed concentration. In the olden days and even these days with great yogis, there is still a practice to observe the disciple for a year and only then impart the ways and means to achieve the state of meditation. Before analyzing meditation let us digress from here. In meditation, the first thing that is achieved is complete relaxation; then, practice withdrawing the senses from outside, which is called Pratyahara; then Dharana, which is an effort to concentrate. During Dharana, concentration goes out and in, just like an electric bulb switched off and on. When concentration is finally obtained, it is called meditation. Meditation is real. Each human being is quite separate in temperament, sensibility, ego and their own karmas embedded in the mind. The same type of meditation will not be suitable for all. When one closes the eyes, the body may not relax immediately. The body may relax for people with suitable temperaments, but majority may not be able to. Similarly, the mind will not remain constantly in the state of concentration either. This will be a problem and we will have to solve it. For relaxing, you will find the following technique quite useful: After sitting in the meditative pose command mentally each of your body part to relax by commanding as follows: big toe relax, second toe relax, middle toe relax, fourth toe relax and little toe relax. Similarly then take, lower sole, inside foot sole, outside leg sole, heels on all sides and top portion of foot then the bone joints of foot lower limb, then take the lower limb on all four sides from lower to new portion. Thus the entire lower limb is relaxed. Then take the knee round on the cap, left side, right side, inner side. Then go on four sides of upper limb on the thighs side and from bottom cover to the hip. Then cover the right buttock on inner sides, outer bottom etc. Similarly start with the left toe and come to hip. Then command the entire hip in a round fashion, relax the genital organ completely by running your mind all over. Thereafter, take the entire stomach as if one is mentally placing the palm over the stomach and asking the entire stomach portion to relax. When this is completed, think about the pancreas, liver, kidney, bladder, intestines and relax them. Mentally command your breasts to relax, left and right. Mentally command the heart to run in a smooth and relaxed way without strain. Cover the left side of the torso from armpit to hip and then the right side of torso also. In the same, do the upper limbs starting from thumb to small finger and upward to the joint in the shoulder. Thus both limbs are covered. Thereafter go to the backside and relax the left portion imagining your palms running across your neck to hip and then the right portion. Here, it would be advisable to run your palms two or three times in different parts of the left back. Similarly on the right back and then run them through the spinal column from neck to bottom. Now, come to the shoulders from the arm joint to neck both left and right one by one. Then shoulder on four sides one by one to the head and chin. Relax the lower chin, then upper chin, lips, inside the mouth, tongue. Tell your teeth to remain strong and not to relax. Further go to the cheeks left and right and then to side portion of the nose left and right and take the ears on all four sides making a round of each ear then the nose, then gradually command the eyes to relax. Go the forehead first on the left and then right and center and finally the entire forehead mentally palming it to relax. Last relax the entire head taking the front portion, back portion, left side and right side. This is a method will take at least ten minutes. Those who are interested to go for easier ways can adopt one of the following two practices: Chant OM while inhaling and HUM while exhaling. Concentrate on your naval portion. With each breath start counting backwards from nine to zero. In the initial state it may be slightly difficult. However, by practice one will be able to achieve it in due time. If you are unable to maintain concentration, results may not come. Practice again. By the time zero is counted the body may relax in due course of practice. Alternatively, concentrate on the breath counting as long as possible or without counting also. Counting while breathing activates a higher state of concentration; when the body is completely relaxed or relaxed to the best of one's capability, meditation can be started. To meditate, one has to concentrate on something. There is no hard and fast rule as to what one has to concentrate on. One can take any God of one's choice or any particular object which is well liked or which is much known to the person. Alternate suggestions from the Yoga point of view is that one of the chakras may be recommended. When relaxation is over meditation automatically has to start. On a given subject try to see through the inner eye before your forehead and try concentrating. The mind will spring from one to another object and many memories will come back. Both the immediate past and previous past in the present life will be shown before one as if you are seeing a film on the screen. When such things come, do not apply your mental ego part on such scenes. Try to be aloof from them and watch impartially. Such impartial watching is as good as unwinding the memories. This will take a long duration of practice to get the mind to a purity level. Some will start seeing a bright light like a sun, some will see thunders and lightning; some will see red hot lines, some will see blue sky, and such many phenomenon's may come before you. Some may see horrible creatures, some may see good saints, and some will see angels, and so on so forth. These are to be observed with dispassionately and without attachment. Meditation of course calms the mind. There is no doubt about this and it has been proven scientifically too. For ordinary human beings, relaxation itself is quite sufficient in today's busy world. Those who are of the intellectual type can progress forward and this will come naturally to them, after few days of practice. In the Yoga Sutras of Patanjali, there are eight parts for realization. These are Yama, Niyama, Asana, Pranayama, Pratyahara, Dharana, Meditation and Samadhi. Yama consists of right attitudes that one has to take in one's life, whereas Niyama are the rules to be observed in life which will influence the body as well as the mind; and in the present western world, Asana means posture, and Pranayama is the control of the breath. The first four parts of yoga as described above are interconnected and if one is practiced, others will be easy to practice. For example, if one practices Asanas and Pranayamas one will automatically gain the effects of Yamas and Niyamas. Similarly, the latter four parts are also interconnected as one leads to the other. There are various Asanas that are in practice. They are all variations of basic 84 Asanas which are well described in Gheranda Samhita. The present day Asanas were not part of the original Yoga practices prescribed in the earlier stages of the seers. They are part of Tantra Shastra which is actually a separate science developed on the body, mind and soul. Many Asanas and Pranayamas combined have therapeutic value which is a separate science in itself. Heavy Asanas are not recommended for those suffering chronic diseases. These Asanas are not recommended especially for those suffering from hypertension. They can practice the set of Pavanamukt Asanas coupled with Marjara Asana and Akarna Dhanur Asana without keeping the breath inside which is called Kumbhaka. Before going any further, it is necessary for one to understand some of the ancient scientific factors related to body, mind and soul. Ancient Vedanta seers and philosophers are of the opinion that Brahma is the one and the only one without a second. Brahma is indescribable with dazzling light and everything emanated from it. It is truth, pure light, intelligence and pleasure. This is to be read on the macro level. There are differences at the micro level, though. One opinion is that after realization, the personal Atman became one with the Brahma. A second opinion says Atman becomes similar to Brahma. Third opinion says Atman is quite distinct. Despite the difference on the matter of the Soul, everybody agrees without dissent that there is Prakriti, the material part from which the world has come out, the world as we see it. This Prakriti the seers say is both truth and untruth. Truth because it remains in macro level and never dissolves except becoming dissolved and remaining dormant in realized people; untruth because it dissolves itself, melting its parts one by one during the practice of yoga in advanced stage. It suffices to understand this much without going into arguments. Prakriti is called Mahat from which Ahankara emanated firstly and this has three attributes called Trigunas. These are Satva (good, light, fluorescence, white, etc.), Rajas (Active and Red) and Tamas (ignorance and black). From the three different attributed parts of Ego emanated the five Tanmatras, the minutest matter considered minuter than the present day concept of Atoms. The combination of these Tanmatras in certain proportions of the three Gunas resulted in Mind, Budhi , personal Ego and Soul. This is the combination of the four parts of the internal organ, the mind. The matters that emanated from the Mahat one by one in descending order are called five elements which are Tanmatras as said above. Akasha, which cannot be equated to the present ether and it is much more than that. From Akasha came Vayu, which is mainly Prana but is air in the physical sense. From air came Agni or fire again to be understood mainly as heat and cold. From Agni came Apas which is water, which is also to be understood as liquidity. And from Agni came Bhu which may be understood as solidity. The entire body structure as well as the mind structure is made of a combination of these five elements, which are minute in nature in certain proportions. For descriptions to understand the body composition, one has to study Ayurveda, a separate science. Similarly for understanding the genes, one has to refer to astrological science. Genes may not be visible for normal persons through astrology. For this, one needs to go deep with intense devotion. There is a saying in the Mythologies of India as follows: "Those destined to know the meaning of Vedas (knowledge) automatically comes to know. Others, despite repeated instructions will never come to know." As such, there is no use delving into Vedas unless one is destined for them. Also, the meaning will be revealed in different aspects and angles to the persons according to their devotion and practice. The meaning thus revealed will be without much variation from main theme. As per ancient science, Agni or the 'heat and cold' is responsible for the conduct of the bodily functions. Vayu, or the air, is responsible for the conduct of the Pranic functions. Aditya, also called Indra i.e. waves of light, thunder, waves of sound etc. is responsible for the egoistic soul of the human being. Everybody knows physiology and therefore we are not delving into this matter. Let us go further. Every science student during his days of study in the school might have seen the microscopic picture of the sperm. This is the major aspect which takes its shape into body keeping itself in the centre portion of the body as the brain and the spine. It is here the entire mechanism of life is centered around and expands out. For Yoga, this particular portion is extremely important. Also important is the energy which regulates the entire day to day operations of the body and the mind. This energy is centered slightly above the root of the spine remaining dormant till it is awakened by yogic practice. As long it is not awakened, it just conducts the normal functions. Once awakened, an enormous power can be experienced by the practitioner. Whatever yoga we practice, awakening of this energy is the primary necessary corollary that needs to be achieved. Pranic energy is actually part of the macro energy which remains dormant in the micro human being and conducts the day to day affairs. There are three energy channels in every human form, although not visible to the ordinary eye. The formation may be compared to a central column and two serpents on both sides of the column. The serpents curve three times around the central column. These three are channels through which energy passes from the bottom emanating from the central energy. This energy's noticeable form is our breath. The left channel is called Ida and the right channel is called Pingala and the centre channel is called Sushumna. The channels are also called Naadis in Sanskrit. This energy has evolved as a combination of various vibrations and energies that are running in the human being and the cosmic universe. They had found that different alphabets when joined together form some other form of energy. Each letter is related to a particular form of wave of energy. This was the basis of Mantra Shashtra. All matter in the minutest form vibrates with expansions and compressions every now and then. Hence, the constant flux of all forces within the body is based on these vibrations which relate to sounds. As such, the Brahma is also called Aksharama Brahma which has two meanings. The first meaning is something which cannot be destroyed. The second meaning is that which emanates from or is connected with syllables, Aksharas. So due to this cause of vibration, two channels as discussed above always vibrate resulting in breath. The seers found this and decided to control their breath. This would control their vibrations. These vibrations due intensive meditation create extreme heat similar to atomic fusion. This control subsequently makes the dormant energy to arise from its sleeping state to awakened state. The energy eventually rises and goes up through the central channel to the Sahasrara chakra in the head. Whatever yoga once practices, the above state must be obtained and only then can one say that progress has been achieved. This happens through intense devotion through concentration or intense chanting of mantra which results in concentration. This results in the awakening. Any type of concentration on a subject in meditation also makes achieves this result. Therefore, the awakening of this prime ordeal energy is the object for realization although ways and means may be different. This is actually called the Kundalini awakening. Before the Kundalini is awakened one has to keep the body and mind very pure as otherwise such awakenings can result in abnormal situations which one cannot control. To avoid such abnormal situations, it is a must that one has a guru. One who has experienced the difficulties during practice or has been taught by a guru alone can help the seeker to avoid such pitfalls. This awakening normally does not happen to everybody but only to those who have been destined. Others drop out during the practice itself automatically discouraged by various shortfalls. Only those who have the grace of the Creator either directly or through their guru can proceed further. Even if those who proceed, they may face financial, mental, physical and various problems which could force them to discontinue their practice. But despite all the problems if one persists in his practice, disregarding these disturbances, they will definitely go towards the goal. As said previously these two channels along with the central channel, are the most important parts to be aimed at in yoga and meditation. The left Ida channel which is actually mental energy has to combine with right Pingala channel energy which is physical. Their combination is preliminary yoga which after much practice results in ultimate yoga of joining the Prakriti with the soul. Dissolving the Prakriti with the soul is called realization. The cleaning of these channels is done by practicing Pranayama and the mind is purified by practicing Yamas and Niyamas and Vasanas. Impressions are cleaned with meditation but karmas cannot be cleaned. One has to live with them but with one's karmas albeit with lesser effect. The physical appearance of life is found to be in the breath of human being. The breath normally flows in either nostril at an interval of ninety minutes. As long as the material parts of the body are in condition, they expand and shrink causing breath. This is due to the energy centralized in the body just above the root and outside of the spinal cord. This is the primordial energy which remains dormant until it is awakened by yoga. As long as it is dormant, only a small percentage of energy works in human beings branching out into two channels as solar energy which flows in the right nostril and the lunar energy in the left nostril. Both these energies travel through the respective channels representing physical and mental energies. The energy flows in the central channel Sushumna four times a day normally. These times are called Sandhyas. The times corresponding to such Sandhyas are early morning when the sun rises, mid day, at the time of sunset and at midnight. Meditation is recommended preferably just before these times for this reason. This does not mean meditation cannot be done in other times. One can practice it as per one's convenience. The flow of energy in the central channel also happens when one is deep into concentration. Similarly, this energy flows in the central channel when one is having sex and at the height of the act. The central channel called Sushumna has again one more Naadi inside which is called Chitra. Chitra engulfs another Naadi called Brahma Naadi. This Brahma Naadi is the exact Naadi on which yogis concentrate. Concentration on Sushumna alone can lead to concentration on

Brahma Naadi. In this central channel are the chakras or the centers of energy. Though some of the recent experiments in Japan and Russia have confirmed the existence of such energy centers, yet they cannot be confronted or seen physically. These are visible only through the eyes of a yogi. In advance meditation practice, for some practitioners these energy centers automatically come into vision in the center of their forehead. In the Tantric texts six such chakras are mentioned. These are: Mooladhara, Swadhishtana, Manipoora, Anahata, Visuddhi, Ajna and above these six there is another one called Sahasrara. Mooladhara is at the end of the spine from where three channels originate. Here the energy starts to flow up. All the basic human natures come out of this place. Swadhishtana is just one inch above the Mooladhara chakra. This is the place for the Ego or the feeling of self which is hence called Swadhishtana. Manipoora is exactly behind the naval in the spinal cord. In this chakra both the mind and soul or the impressions and vasanas are installed. Anahata is in the center of the heart slightly at the right side and it is here the feelings and sensations are situated. Visuddhi is at the place of the thyroids below the neck. There is a special Naadi just below this chakra which is responsible for dreaming. Ajna is in the centre of the forehead in between the two eyes. This charka has the capacity to command all the functions and hence is called Ajna. During meditation if there is no peculiar desire to concentrate on any other centre, it is better to concentrate on this centre rather than coming up concentrating from Mooladhara up. The Sahasrara chakra is above all the other chakras. It is at this chakra that one gets the nirvana or realization or enlightenment. The notion of chakras as centers of consciousness is a little more compelling than simply calling them an "energy nexus," which is largely meaningless in context. It's probably not coincidental that the chakras correspond to physical locations where people commonly describe holding sensations or emotions, such as a broken heart, a gut instinct or words that catch in your throat. As an added bonus, there is currently no way to scientifically measure consciousness. The philosophy of yoga is to help one achieve a certain poise of equanimity, to look at all aspects of life—pleasure as well as pain—with acceptance and serenity. And although yoga's roots lie deep in the Hindu world of gods and goddesses, yoga itself is not a religion. Religion, one can argue, is often about teaching you what to do. Yoga teaches you how to be. The fact is, yoga is universal and practitioners are always welcome to use their own concept of God to help guide them toward higher levels of yogic performance. Besides, tenets of yoga are found in many of the world's major religions. For instance, Christendom's St. Francis of Assisi was an avowed yogi. And Judaism's Kabbalah draws heavily on aspects of the yogic science of breath. Despite our own true nature, people usually become far too distracted with their own mind and body and material objects. They lose sight of this fundamental truth. This false identification makes us feel imperfect, limited, sorrowful and at loss. Yoga seeks to provide people with a way to cast off this ignorance and become aware of their true divine self. The goal is to free a person from those imperfections and to unite him or her with their supreme universal self. Yoga, then, is the union of the individual with absolute reality—and the yogi their journey is all about finding peace, harmony and the greatest truth of all. Learning to meditate is the greatest gift you can give yourself in this life. For it is only through meditation that you can undertake the journey to discover your true nature, and so find the stability and confidence you will need to live, and die, well. Meditation is the road to enlightenment. Through meditation we are able to integrate all the illusions of separation, and merge our awareness into the intuitive state where we realize the underlying infinite, conscious principle that is both the very essence of all.

The Chakras

It is said that when you begin to develop your senses, a new and fascinating world opens before you; the hidden world suddenly unveils itself - your perception heightens and your thoughts and feelings are expressed before your very own eyes in color and form. There is more to the human body than the physical body. Unfortunately, most people consider the physical body and the material world to be the only reality that exists. They believe this because for them, these are the only things that can be discerned with their own physical senses, and I might add, understood by their rational mind. But there are numerous energy bodies within and around the human body. These energy bodies are: the ethereal body, the emotional (astral) body, the mental body, and the spiritual body. Each energy body possesses its own vibrational frequency, from the lowest (ethereal) to the highest (spiritual). In addition, there is a complex energy system that is at work, which the body could not exist without. This energy system correlates with the acupuncture points and consist of energy bodies, namely: the chakras or energy centers and the Naadis (also known as energy channels). Naadi is a Sanskrit word meaning, pipe or vein. Naadis are akin to a network of channels or arteries that transport vital energy throughout the human being's energy system. Chakras are energy centers. They function like receivers and transformers of the various forms of prana. Through the Naadis, the chakras take in the vital energy and transform it into the frequencies needed by the various areas of the physical bodies for sustenance and development. Each chakra is connected with one of the elements of earth, water, air, ether and mind - mind being an instrument of consciousness. These elements are states of matter and not elements as we understand them in modern chemistry. They are equivalent to the terms: solid, liquid, fiery or gaseous, airy, and etheric - which are somewhat analogous to the physical, astral and mental planes and sub-planes. Chakra is a Sanskrit word, meaning, wheel. The human chakras are wheel-like vortices, or saucer-shaped depressions that exist on the surface of the etheric human body. Traditional writings say there are 88,000 chakras in the human body. Most are extremely small and play a minor role in your energy system. However, there are approximately forty secondary chakras that are of significance; these are located in your spleen, the back of your neck, the palms of your hands and the soles of your feet. But for brevity purposes, we will only explore the seven primary chakras here. The first chakra is called the Root chakra. It is located between the anus and the genitals and is connected with the coccyx. The Root chakra opens downward. When active with vigor, it is fiery red-orange in color. It is the level of energy- consciousness we use to manifest our body on the physical plane of existence, so that our spirit may be able to have experience in the physical realm. The energies here are concerned with survival issues and in the unbalanced state, manifest as fear, selfishness, possessiveness, etc. It represents dormant; coiled up as a serpent and uncontrolled sexual energy. The task is to convert this sexual energy into spiritual energy, thereby achieving expanded consciousness in the realms above and beyond the physical. The second chakra is called the Sacral or Spleen chakra. Located over the spleen, it is sun-like in color and opens towards the front. Sanskrit books always substitute the second sacral chakra with that of the spleen, locating it below the navel instead of at the spleen. My research suggests that there is a danger associated with tampering with the spleen, so perhaps this is why they locate it below the navel. It is the area of the organs of procreation where, if controlled by the higher centers, cosmic energy may be absorbed and transmuted to create more than just physical offspring; one may give birth to their untapped creative talents. However, if the individual is psychologically developed to this level only, he or she will be preoccupied with sexual matters. The third chakra is called the Solar Plexus chakra. It is located about two fingers above the navel and is directly connected to our astral or emotional body. Through the Solar Plexus chakra we absorb the solar energy which nurtures our etheric body, which energizes and maintains our physical body. This is where our emotional (feeling) energy radiates, particularly our gut feelings. Its color is chiefly green. The vital fire which sustains the physical body is located here. This energy center gives one a sense of power and will. Meditation on this spot sends vital energy to this center which promotes health by increased digestion and equal distribution of energy throughout the body; the effect is physical and emotional balance. This is a prerequisite for higher spiritual advancement. Imbalance or obstruction in this energy center manifests as feelings of inadequacy and powerlessness; one feel incapable of expressing their willpower. The fourth to the seventh energy-consciousness centers (chakras) represent the evolution of spiritual life, the divine nature which is latent in everyone. The lower energy centers (three to one) represent the life of the lower nature, the animal nature which seeks survival and supremacy over others and gratification of the senses. Moving from the lower to the higher consciousness represents re-birth, the birth of the spirit. The fourth chakra is called the Heart chakra. It is the center of our entire chakra system. It is located in the center of the breast at the level and vicinity of the heart cavity and connects the three lower physical and emotional centers to the three higher mental and spiritual centers. It glows a golden color. This center opens up the possibility of the virgin birth which is the spiritual consciousness as opposed to the lower animal consciousness. At this level, your spiritual essence becomes revealed in the caring; expressing love for others. This chakra is related to a gland in the body known as the thymus. The thymus is intricately related to the immune system, producing antibodies to ward off infections. The fifth chakra is called the Throat chakra. It is located between the depression in the neck and the larynx, beginning at the cervical vertebra behind the Adam's apple. It starts at the cervical vertebrae and opens towards the front. It is also connected to a small secondary chakra, which has its seat in the neck and opens to the back, but since the two chakras are so closely related, they have been integrated into one. Although it has a lot of blue about it, it is also silvery, like moonlight upon rippling water. Blue and green predominate. This is the chakra of divine speech. This center is involved with our ability to receive love from others; our self value. Imbalance in this energy center may be expressed as the inability to communicate our needs to people around us and the world at large. The sixth chakra is called the Third Eye chakra. It is associated with the pituitary gland, which is a very small, shapeless organ about 1/8 inch in diameter located in the forehead about one finger-breath above the bridge of the nose between the eyebrows. Here conscious perception of being takes place. It is the seat of our higher mental powers. On the physical plane, it is the highest center of command for the central nervous system. One half is chiefly rose-colored with a lot of yellow, and the other half is predominantly purplish-blue. It is the level of intuitive wisdom that does not require thinking or the processing of the mind; the light if intuitive vision. It is the level of knowing something instead of thinking we know something. Herein lays the ability to perceive spirit as well as matter, and to live in both equally. The seventh chakra is called the Crown chakra. It is seated in the pineal gland, which is a small organ of fleshy consistency not much larger than the pituitary. The pineal gland is located near and behind the pituitary body almost in the exact center of your head at the level of your ears. The Crown chakra opens upward, at the top of your head. Interestingly enough, medical science has yet to conclusively determine the physical influence this gland has on the human body; probably because metaphysics is beyond their rational thinking mind. Although it contains all sorts of prismatic hues, it is predominantly violet. This is the energy-consciousness level where the dualistic mind and the life giving fire (primal energies), becomes united into one whole. This represents the level of transcendence, where the anointing occurs. It is here where the merging of the individual soul with the absolute reality occurs. Together the seven chakras form a system for modeling consciousness that enables us to better see ourselves in mind, body, behavior, and culture. This system is a valuable tool for personal and planetary growth. As all our action and understanding arise from and return to points within ourselves, our chakras, as core centers, form the coordinating network of our complicated mind/body system. From instinctual behavior to consciously planned strategies, from emotions to artistic creation, the chakras are the master programs that govern our life, loves, and learning. Like a rainbow bridge, they form the connecting channel between mind and body, spirit and matter, past and future, heaven and earth. The chakras are the gears that turn the spiral of evolution, drawing our attention toward the still untapped frontiers of consciousness and its infinite potential. Chakras exist in many dimensions simultaneously, and as such, are a point of entry into those dimensions. In the realm of the physical, they are the areas of the body and can be measured as patterns of electromagnetic activity, centered around the major nerve ganglia. Butterflies in our stomach, frogs in our throat, pounding in our heart, or the experience of orgasm; these are all manifestations of the presence of chakras in your body. In the world of the mind, chakras are patterns of consciousness; belief systems through which we experience and create our world around us. The one who sees the world in terms of love and compassion will create that reality by her belief system. As we change our paradigms of consciousness, the world around us changes, both individually and collectively. Our belief systems, at the core of our being, create the chakras which govern our lives. These are the belief with which we create our world. In all dimensions, chakras are a kind of vortex; a significant gathering point of organized life energy. The vortex is a gateway connecting each of these dimensions through one central entity or concept. The gateway exists within ourselves; through it we have access to all we could wish to explore. When all of the chakras are understood, opened, and connected together, we have bridged the gulf between matter and consciousness, understanding that we, ourselves, are the bridge. The chakra system, with its seven levels, is not new by any means. It has been a spiritual guide for mystics and yogis for at least five thousand years, with its origins buried in the roots of Vedanta philosophy. They say the earliest mention of the word chakra in written literature is in the Vedas, where Vishnu, a major Indian male deity, is described as descending to earth, having in his four arms a chakra, a lotus flower, a club and a conch shell. The actual age of the Vedas has been a subject of debate to scholars; most agreement seems to center around 2,000-600 B.C. However, yogic systems which systemized the chakras goes back even earlier than pre-Vedic and pre-Aryan times to Sumer, a civilization and historical region located in Mesopotamia, known as the cradle of civilization (5300 – 1940 B.C). Descriptions of the chakras as meditational centers appeared in some of the later

Upanishads, briefly in the Yogas Sutras of Patanjali, and later thoroughly translated into a popular and standard text 'The Serpent Power', written in the 1920s. Aside from this literature, systems involving seven levels of man, nature, or physical planes are quite common. The theosophists, for example, talk about seven cosmic rays of creation, with seven evolutionary races. The Christians talk about seven days of creation, as well as seven seals, seven angels, seven virtues, seven deadly sins, and in Revelation 1:16 perhaps even seven chakras where it is said, "And he had in his right hand seven stars." Seven-ness is also found outside of myth and religion. There are seven colors in the rainbow, seven notes in the Western major scale, seven days in a week, and it is believed that major life cycles run in periods of seven years each; childhood to age seven, adolescence at fourteen, adulthood at twenty-one, etc. Many cultures talk about energy centers or levels of consciousness similar to chakras, although there may not always be seven centers in their system. The Chinese have a system of six levels in the hexagrams of the I Ching, based on the two cosmic forces, Yin and Yang. There are then six pairs of organ meridians which correspond to five elements which was discussed earlier. Fortunately, more and more research is being done now that supports the existence of chakras, and their counterpart, the kundalini energy. The system works as a theoretical model of consciousness, dealing with an expansion in awareness that can only be felt through trial and error. So, in order to understand the merits of this most ancient and now modernized system, I urge you to suspend disbelief within whatever parameters you find comfortable, jump aboard the mystic bandwagon of personal experience, and judge their truths from within.

Pranayama

Ether is the all-penetrating existence. Everything that has form, everything that is the result of combination, is evolved out of ether. It is ether that becomes the air, that becomes the liquids, that becomes the solids; it is ether hat becomes the sun, the earth, the moon, the stars, the comets; it is ether that becomes the human body, the animal body, the plants, every form that we see, everything that can be sensed, everything that exists. Just as ether is the infinite, omnipresent material of this universe, so is prana the infinite, omnipresent manifesting power of this universe. At the beginning and at the end of a cycle all tangible objects resolve back into ether, and all the forces in the universe resolve back into prana. In the next cycle, out of this prana is evolved everything that we call energy, everything that we call force. It is prana that is manifesting as gravitation, as magnetism. It is prana that is manifesting as the actions of the body, as the nerve currents, as thought-force. From thought down to physical force, everything is but the manifestation of prana. Pranayama is the control of prana. Suppose, for instance, a man understood prana perfectly and could control it, what power on earth would not be his or hers? Thus, all books and philosophies have been written for the purpose of demonstrating that one thing by the knowing of which everything is known. I'm sure you are curious how a man or wombman can know all through knowing particulars? Behind all particular ideas stands a generalized, abstract principle. Grasp it and you have grasped everything. He who has controlled prana has controlled his own mind and all the minds that exist. This body is near to us, nearer than anything in the external universe, and the mind is nearer than the body. But prana which is working this mind and body is the nearest. It is a part of the prana that moves the universe. The vital force in every being is prana. Thought is the finest and highest manifestation of this prana. Conscious thought, again as we see it, it is not the whole thought. There is also what we call instinct, or unconscious thought, the lowest plane of thought. All reflex actions of the body belong to this plane of thought. There is again, the other plane of thought, the conscious: I reason, I judge, I think, I see the pros and cons of certain things. Yet that is not all. The circle within which it runs is very, very limited indeed. Yet at the same time we find that facts rush into this circle, like comets coming into the earth's orbit. It is plain that they come from outside, although our reason cannot go beyond. In this universe there is one continuous substance on every plane of existence. Physically this universe is one: there is no difference between the sun and you. There is no real difference between this book and you: the book is one point in the mass of matter, and you're another point. Each form represents, as it were, one whirlpool in the infinite ocean of matter. There is always constant change. Not one body remains the same. There is no such thing as my body or your body, except in words. Matter is represented by the ether, when the action of prana is most subtle, this ether, in a finer state of vibration, will represent the mind, and there it will be still one unbroken mass. The Universe is an ocean of thought, and we have become little thought whirlpools. Thus even in the universe of thought we find unity; and at last, when we get to the self, we know that the self can only be one. Beyond the vibration of matter in its gross and subtle forms, beyond motion, there is but one. The sum total of the energies in the universe is the same throughout. It has been proven that this energy exists in two forms. It becomes potential, unmanifested, and next it becomes manifested as all these various forces; again it goes back to the quiet state, and again manifests. Thus it goes on evolving and becoming involved throughout eternity. The control of this prana, as before noted, is what is called Pranayama; which has very little to do with breathing. But the control of the breath is a means to the real practice of Pranayama. The most obvious manifestation of prana in the human body is the motion of the lungs, which is associated with the breath. Pranayama is not breathing, but controlling the muscular power which moves the lungs. That muscular power which is transmitted through the nerves to the muscles and from them to the lungs, making them move in a certain manner, is the prana we have to control through the practice of Pranayama. Wherever there is life, the storehouse of infinite energy is behind it; starting as some very minute, microscopic bubble, and all the time drawing from that infinite storehouse of energy, a form changes slowly and steadily, until, in the course of time, it becomes a plant, then an animal, then a man, and ultimately God. This is attained through millions of eons, but what is time? An increase of speed, an increase of struggle, is able to bridge the gulf of time. All the great prophets, saints, and seers of the world- what did they do? In one span of life they lived the whole life of humanity, transverse the whole length of time that it takes an ordinary man to come to perfection. In one life they perfected themselves; they had no thought for anything else, never lived a moment for any other idea, and thus the way was shortened for them. This is what is meant by concentration: intensifying the power of assimilation, thus shortening time. We have five senses, actually nine but I won't go into that, and we represent prana in a certain state of vibration. All beings in the same state of vibration will see one another; but if there are beings who represent prana in a higher state of vibration, they will not be seen. Our range of vision is only one plane of the vibrations of prana. Take the atmosphere, for instance: it is piled up layer on layer, but the layers nearer the earth are denser than those above, and as you go higher the atmosphere becomes finer and finer. Think of the whole universe as an ocean of ether, vibrating under the action of prana and consisting of plane after plane of varying degrees of vibration. Visualize it as a circle, the centre, the slower are the vibrations. Matter is the outermost plane; next comes mind; and spirit is the centre. We are all parts of the same ocean of prana and we differ only in the rate of vibration. Now, there are two nerve currents in the spinal column called the

Pingala and the Ida, and a hollow canal called the Sushumna running through the spinal cord. The left side is the Ida, the right is the Pingala, and that hollow canal which runs through the centre of the spinal cord is the Sushumna. We know that there are two sorts of actions in the nerve currents: one afferent, and the other efferent; one sensory, and the other motor; one centripetal, and the other centrifugal. One carries the sensations to the brain, and the other, from the brain to the outer parts of the body. The spinal cord, and the brain, ends in a sort of bulb in the medulla, which is not attached to the brain but floats in a fluid in the brain, so that if there is a blow on the head the force of that blow will be dissipated in the field and will not hurt the bulb. Secondly, we have also to remember that, of all the centres, three are particularly important: the Muladhara (the basic), the Sahasrara (the thousand-petaled lotus, and the Manipura (the lotus at the navel). Next, is the correlation of electricity. What electricity is some know, but so far as it is known, it is a sort of motion. Electricity becomes manifest when the molecules of a body move in the same direction. Another point we must remember is that the nerve centre which regulates the respiratory system, the breathing system, has a controlling action over the whole system of nerve currents. Hopefully you are capable of seeing why rhythmical breathing is practiced. In the first place, from it comes a tendency of all the molecules in the body to move in the same direction. Therefore when all the motions of the body have become perfectly rhythmical, the body becomes a gigantic battery of will. This is, therefore, the physiological explanation of Pranayama; it tends to bring a rhythmic action in the body, and helps us, through the respiratory centre, to control the other centres. The aim of Pranayama is to rouse the coiled up power in the Muladhara, called the Kundalini. Thus, the rousing of the kundalini is the one and only way to the attaining of divine wisdom, super conscious perception, realization of the spirit. There are various types of Pranayama which belong to Tantra Shastra. Nadishuddhi Pranayama is the one mainly mentioned in the original Yoga sutras and other Pranayamas are Bhramari, Sheetali, Sheetkari, Ujjayi, Kapalabhati, Moorcha, Plavani, Suryabhedi, Chandrabhedi, Bastrika and others. Such pranayamas are recommended for advanced practitioners while for normal meditational purposes Nadisuddhi alone is sufficient. But this practice has to be improved by having retention of breath preferably either inside or outside or both. This retention is called Kumbhaka.

It has to be practiced by taking breath inside and close both nostrils and keep the breath inside to a comfortable level so that exhaling will be smooth. In the initial stages, exhaling can be sudden. By practicing gradually one will become an expert. The initial practice of Pranayama can be started in the ratio of 4 numbers counted for inhaling, 4 numbers for retention and 4 numbers for exhaling. Then increase next week to 4 numbers of inhaling, 6 numbers of retention and 6 numbers for exhaling. Next week 4 numbers inhaling, 8 numbers retention, 6 six numbers exhaling and in this way, ultimately bring it to 4:16:8 ratio. Thereafter go for 5:20:10 and then 6:24:12. Beyond this ratio it is not necessary for normal practitioners. There is no harm if ratio is varied as per capacity. Once internal retention is practiced and perfected, external retention should be practiced. The count number should be half of the internal retention. In meditation, regular practice will make one perfect within six months to one year. The breath which goes like a serpent will come under control and the accompanying rash sounds will vanish; length of the breath will increase and become very minute and will not be heard even by the practitioner. Body will become lighter, fat will reduce, Obesity will go, and the mind becomes calm. Pleasure will be felt and every action done by one will become without strain. Eyes will become bright and ears become very sensitive. Better tastes will be felt and addictions will go. Color will change and faces get uplifted and there are many benefits that come from the Pranayama practice. In addition, Pranayama can hasten the development of mind purity with the retention and distribution of prana in the body. It is very difficult to concentrate the mind and attach it to one place only; the mind is a slave of physical objects. There is hardly a child born who has not been told, "Do not steal," "Don't tell a lie"; but nobody tells the child how they can avoid stealing or lying. Only when we teach our children how to control their mind do we really help them. The goal of each soul is freedom, mastery; freedom from the slavery of matter and thought, mastery of the external and internal nature. First hear, then understand, and then, leaving all distractions, shut your mind to outside influences and devote yourself to developing the truth within you; to succeed you must have tremendous perseverance, tremendous will.

Quantum Reality

Humanity's first prototype began as long-lived, god-like, ethereal, hermaphroditic beings, Etherian Sirians, who gradually polarized into opposite sexes and solidified into melanin; black flesh form. Melanin is the chemical key to life itself; it is essentially linked to the DNA of the genes. Therefore, melanin is essential for reproduction. Blackness is a divine, cosmic principle of the Universe. Black is the meaning of Kam (Ham, Khem) - the name which the ancient Egiptians called themselves. From this word we get chem-istry, which means, the study of blackness. Life is founded upon carbon, which is the sixth element on the periodical chart; it has six electrons, six neutrons and six protons; the element present in all living matter. Carbon atoms link to form melanin, which has black hole properties. Black holes are found at the center of our own galaxy and countless others. In physics, a black body is known to be a perfect absorber and perfect radiator of all forms of light waves, which are energy particles or frequencies, such as, electrical, infrared, visible [sun-light - red, orange, yellow, green, blue, indigo, violet], ultra violet, x-rays, music, radar, etc.; can transmute (from negative to positive particles) and store this energy for later use. This black body radiation is at work in the electron, as shown by Nobel Prize winner, Richard Feynman. The electron is responsible for all chemical changes in matter. It has been present since the creation of the universe. The scientist, Jean Charaon proved that the electron has all the properties of the black hole, plus it exchanges black photons with other electrons, enabling it to continuously accumulate data. This means that if we view the electron as a carrier of memory, it has experienced everything in creation, since the very beginning. Blackness is fundamental to the operation of the universe of energy. The Creator is the giver of all energy: Black allows the perfect reception of all wavelengths; "Blackness" ("Triple State of Blackness" or "Triple Stage Darkness") is a divine, cosmic principle of the Universe. Dr. Welsing writes in <u>"THE ISIS PAPERS"</u>

> "Since melanin is a superior absorber of all energy, it is essential to establish this understanding of God and 'ALL energy...Melanin gives us the ability to use our bodies as direct connections with the God Force, the source of all Energy-like plugging a cord into a wall socket."

Carol Barnes writes "Melanin is responsible for the existence of civilization, philosophy, religion, truth, justice, and righteousness." Melanin is excreted for the Pineal Gland (approximately the size of a pea, hence the term "Pea Brain"), in two forms during the hours of 11pm - 7am, melatonin (mellow) is produced and between the hours of 7am - 11pm, serotonin (serious) is produced in the hue-man body. Originally, the pineal gland was the size of a quarter and we (hue-mans) were able to breathe (meaning we used to bring prana) directly, into the top of our heads; illuminating the "master gland". However, scientific research reveals that most European people are unable to produce much melatonin and serotonin because their pineal glands are often calcified and non-functioning by the age of seven. The pineal gland calcification rates among Moors are 5-15%; Asians - 15-25%; Europeans -60-80%! There are several healing sounds that can decalcify the pineal gland. They are the sounds of healing, 'MAY', 'THOH', 'AYIN', 'I', 'E'. Information is carried to and from the brain and spinal cord by axons, long thin extensions of nerve cells. Axons group together in bundles which outside the brain and spinal cord are called nerves. Nerves leaving the spinal cord are called spinal nerves, and those leaving the brain are called cranial nerves. These nerves when enlightened give one E.S.P. (Extra Sensory Perception) or H.S.P. (Higher Sensory Powers). The cranial nerves; 12 pairs of nerves come directly out of the brain, and are called the cranial nerves they sit in a circle around the pineal gland, called the master gland of the body, like the external sun as it travels through the twelve Zodiac Signs; Yashua and his 12 Disciples at the Last Supper and King Arthur and his 12 Knights at the Round Table. The pineal gland in the occult teachings is called the "Third Eye." It is actually, the "First Eye." In Hebrew and Arabic, the letter Ayin's symbol is a single eye. Ayin is, therefore, the symbol of perception and insight, of the physical eye and of the spiritual eye. It illustrates symbolically that the pineal gland eye can be taken as a microcosm of the Universe. Ayin or Eye, "wellspring," source, or center and is linked with the Egiptian eye of Horus (Heru [Egiptian], UriEL [Hebrew], whose Latin name is Lucifer, the Light Bearer). In the Tarot Deck the Devil's Card letter is the Hebrew letter Ayin and its numeral is 15, which also symbolizes the Zodiac Sign Capricorn, which is Baphomet, a misnomer of the name Mohamet or Mu-Ham-Mad, the Kama'atu meaning "Water of the Black Truth," in other words "Melanin." Mu-Ham-Mad also symbolize the "Crown chakra." Do you overstand the symbolism that has been placed before you throughout the world? We need to study, and in doing so your perspectives on life with surely change, or rather it will evolve. We came from the stars. We are, literally, made of atoms created and blown into space by ancient stars, a fact that is only one strand in a network that connects us with the rest of the universe. Atoms and molecules are the microscopic constituents of matter. Everything is made of imperceptibly small particles, which explains and deeply influences our culture's perception of reality. An atom is the smallest physical part of a chemical element. Think of all the different material substances around you: this book, your clothes, you hair, etc. You can transform most substances into other simpler substances which is called a chemical decomposition of the original substance. For example, by passing electricity through water, it can be decomposed into two distinct substances, called hydrogen and oxygen. There are 99 elements that cannot be chemically decomposed, and they are called chemical elements, or simply elements. These are the foundation of all other substances; those made of more than one element, called chemical compounds. Imagine dividing a cup of water into smaller and smaller amounts.

If the water is pure, you will always get just water. Every pure chemical compound is made of identical tiny particles that are themselves made of two or more atoms attached together into a single identifiable unit. Such a particle, the smallest compound that has the characteristics of that compound, is called a molecule.

> *A human being is a part of the whole, called by us "Universe," a part limited in time and space. He experiences himself, his thoughts and feelings, as something separated from the rest- a kind of optical delusion of his consciousness. This delusion is a kind of prison for us, restricting us to our personal desires and to affection for a few persons nearest to us. Our task must be to free ourselves from this prison by widening our circle of compassion to embrace all living creatures and the whole nature in its beauty. Nobody is able to achieve this completely, but the striving for such achievement is in itself a part of the liberation and a foundation for inner security...*
>
> *- Albert Einstein*

When it comes to overstanding where you are, and by that I mean in the Universe, we need to establish the concept that there are eleven planes or realms. We reside in the material plane, the densest form of matter which in itself subdivides into different forms of matter based on the motion of the electrons and protons around the nucleus of the atom. For example, iron has an atomic number of 26 which is less dense than carbon which has an atomic number of 6. So the motion of the atom is different; yet it's still matter. Your body is composed of several forms of matter, your teeth are one, your skin is another, your hair, etc. In the physical plane of matter there are various degrees of matter. Once you step out of the physical plane you go into the plane of force. The plane of force stimulates the physical. That means that the trees and the wood are one degree of the physical plane. The life force, the water and the sun and the minor particles of living energy that makes it grow and blossom; that energy is called the plane of force. When you move above the plane of force you go to the spiritual plane. Now you're moving in what is called an etheric state, an exoplasma. You're moving into energies that do exist but until recently science could not explain because the farthest they got was to the hydrogen atom. Then as science became more advanced with technology and computers, they've found out that the atom is composed of smaller particles. The problem starts when you use the word some versus sum. Phonetically they sound the same because they are the same. Something is the sum of a thing; its weight, its mass, etc. When you say a hydrogen atom you're defining, before recently, the lightest sum or thing in existence was hydrogen. When they split the atom, they got two things which became a bi-aps; which simply means two apertures. They don't know what to do or say because for the last four hundred years they've thought that the lightest thing in existence is the hydrogen atom. Now when they flipped the atom they got a quark, which is a quad, or four. So they would had to have split two atoms to get to a quark, so they had to get through bi-aps before they got to quark and before they got to the zeles; which is nothing but an Arabic word meaning to add on. They got passed the first sum, the lightest atom and now they have a second degree of energy and matter which they don't have an atomic weight for. So this verifies that there does exist a form of living intelligence and force that has no sum on this side of things, which is the spiritual world. They just don't want to say spiritual because the next question would be can we communicate with it? Can it help me? Yes. We've been doing that for years; contacting the spiritual world, until Christianity told us to stop doing it and only go through their image of Yashua, which cut off our connection to the spiritual world. So you have the physical plane that subdivides into different form of matter. The plane of force is the energy that is inside all living things. Then the spiritual plane, beneath the atom, a higher level of energy and then you go into the mental plane; mind, the individual portion that feeds off the mental. The mental is the reservoir of intellect of which all things receive. It moves through the thought pattern of the soul, or the spirit. Past the mental plane is the fourth level, or the four elements that manifest the physical. The next level is the plane of divine truth. The truth is not always a reality. Just like the truth was when it was taught that hydrogen was the lightest atom. The next level is reality, but it's not. It was accepted and believed to be true during the period they were not intelligent enough to find out that it wasn't the truth. The plane of truth is what you call yourself. What you are, who you are, your personality, what you've shaped yourself to be. What you want people to think you are, what you really know you are, and what you don't like about yourself; that's the truth. Then there is reality, what are you really; the answer to that manifest only to you and nobody else. You can tell me the truth about yourself, but the reality about you only you admit to cause you can't lie to yourself. So you move from divine truth to divine reality. When you reach divine reality you're at the threshold of the bosom of godliness; where no lies come in, because everything is a reality. As you step from the bosom of god and look back through the seven planes you're looking through the face of god. You're looking from the mask of god, you see less than thyself, so therefore the next step would be higher than myself; another reality, because god is a total reality as the hydrogen was until it was flipped. Now god is a reality, until god says it has to have an origin, point of existence. When you get to the point when you are able to question; you wear the mask of god. God is a mask, you are working your way towards this mask and when you get there you put the mask of god on and look back at creation. Growth, that's the truth but beyond that truth; there's something else that god sees. He sees the only real reality, change. Growth is change; the only real consistency is change, that's reality. That's when you put on the mask of god and look back at creation and creation becomes growth. It's no longer birth, its growth. It's no longer creation, its growth. Then you realize growth is growing toward you, through you, and past you. It's constantly changing. You're born you grow to old age, you die and your flesh becomes a part of soil to grow trees, etc. and it continues to go on. It starts to sum up a second degree of reality. You go four steps. You step from god at 7 to the 8th point, the reality of change; 8 is the number that keeps multiplying by itself. Above 8 is the perfect cycle at 9. That's when all realities meet. Once you get to 9 reality is gone, the quark is gone, the soul is gone, and the higher purpose manifest itself as naught; nonexistence, cause 10 does not exist. 10 takes you back to a new number 1, which is another dimension starting nine more planes of energy, but energy unlike the energy we know. That's why they call it anti-matter. It's on the flip-side of matter. What is interesting about this is in order for one to calibrate anything, you have to be somewhere beside it. You can't add it up without stepping away otherwise you'll be part of the measurement and you wouldn't be able to separate yourself. So you must step away to add up, so when you get to 9 to start the new 1 the anti-matter is on the other side of the forms of matter that we know. So on the other side of spirit there's an evil spirit. On the other side of good there's a bad and you start your next cycle upward, but change stay consistent. The only thing that is definite, constant and sure is change. When you put on the mask and become god you have to become responsible; now you must respond to the needs of humanity and when you are capable of responding to lies and to the historical errors, you are then a god. In conclusion, the basis for quantum reality or quantum mechanics is the recognition that everything has a wavelike nature, even those things we normally consider particles. By the same token, those things that we usually consider waves (e.g., light) also have a particle nature. The Four Fundamental Forces of Nature played a role in both the creation of matter and anti-matter, and the evolution of four of the seven senses of man. It should be now crystal clear that electromagnetic force was the first of the fundamental forces to develop when God electromagnetically recalled all the thought energy that had been scattered throughout the Universe. However, in electromagnetism, like charges repel each other. Therefore, without a strong force of nuclear attraction, two proton could not exist in the same nucleus to form atoms without flying apart. Thus the Universe then became consciously aware of the strong force of nuclear attraction that existed within the tiny black holes scattered throughout the Universe. These tiny black hole functioned like miniature but very deep space warps that empowered the Mind of God to be omnipresent and omnibus throughout the entire Universe no matter how large or how small it was perceived to be. These space warps also produced time warps that empowered the Mind of God to overcome the perception on linear time and space—theoretically creating hyperspace.

Thereafter, all waves that entered into the event horizon of these deep space warps cease to travel in a linear motion and entered into a spin, and thereby took on the behavior and characteristic of particles and anti-particles. This process transformed the oscillation of a wave into the oscillation of a particle. In other words, it transformed the alternating frequency of waves along a linear line to the vibratory frequency of a particles within a certain controlled space or controlled orbit. In Genesis 1:2, these deep space warps are described as the face of the deep. The waves that moved over them are described as the Spirit of God (the thoughts and thought energy of God) moving upon the face of the waters, as in waves of water. Heru described these deep space warps as the watery abyss of Creation called Nu, who represented the fathering behavior of the strong nuclear force of attraction therein. Once all the thoughts of God were coded by the colors and anti-colors carried by quarks and anti-quarks, and sealed in memory by the creation of the neutron, the universe became consciously aware of the need to radiate the stored knowledge throughout itself as life and intelligence—giving everything a purpose, reason for existing. Thus the Universe also became consciously aware of the weak nuclear force associated with radiation. Finally strong gravitational forces were produced that employed the other three fundamental forces of nature to produce heavier and heavier atoms and chemical elements in a gravitationally controlled space. This is essential to perpetuating and sustaining both life and intelligence in a physical and natural existence. However, I have thus far said little or nothing about gravity, or any waves or particles known to carry the gravitational force. Therefore, lets take a brief look at gravity and the possible existence of quantum gravity, and a particle that would contain or mediate its quanta of energy. In modern physics theories, forces are carried by particles. An example is the photon, a massless particle that carries the force of electromagnetism. The graviton is the name given to the particle that would carry the force of gravity. It has never been detected, but is required if gravity is to be understood in the same way as the other fundamental forces. This way of thinking about the nature of gravity is different from the geometrical description in Einstein's general relativity. The two points of view are not necessarily incompatible, but reconciling them (in detail) will require a more complete theory of gravity. As we will see, reconciling a theory of quantum gravity with Einstein's geometrical description of gravity will also reconcile the quantum mechanics' version of spin with the classical mechanics' version of spin. It is all about understanding what the Mind of God was thinking about and what it was trying to accomplish that caused both a classical mechanics version and and quantum mechanics version of gravity and spin to exist. Both the case of gravity and the case of spin developed as a result of the Mind of God working to overcome the perception of time and space necessary to become Omnipresent and Omnibus, as well as to remain Omniscience. Remember, the energy of God's thoughts traveled in oscillating waves on a linear line in to infinity—creating the perception of time and space. During Planck Epoch, the universe was sufficiently dense and energetic to the point that all the fundamental forces, including gravity, were merged into one grand force. Quantum mechanics is generally associated only with the world of the very small, but during the Planck epoch the entire observable universe was tiny. Under such conditions, quantum mechanics and gravity must merge into quantum gravity. Unfortunately, at this time physicists do not have a theory of quantum gravity. To develop a theory of quantum gravity, we must take a second look at the student in a class, wherein the minds of students are bombarded with information transmitted from the mind of that teacher. Remember, the waves carrying God's thoughts in the form of thought energy were sent into a spin in efforts to organize those thoughts and energy while committing it all to memory. Likewise, the mind of the student was also sent into a spin in an effort to organize the information transmitted from the mind of the teacher while committing to memory as much as possible. However, there was something else that occurred before the student's mind was sent into a spend. The first thing to occur involves the information transmitted by the mind of the teacher gravitating towards the mind of the student. As in the meeting of two minds, we therein have two dimensions being drawn to a single point of interest: the dimension of the teacher and all the other dimensions therein, and the dimension of the student and all the dimensions therein also. Though this help to close the gap between the mind of the teacher and the mind of teacher produced by the perception of space between the two, it does not the gap produced by the perception of time. Because the transmission of this information does not occur at once, but rather over a period of time that constitute the span of the lecture, the information gravitates towards the student in quanta of information. Once all the quanta of information are absorbed by the student, the gap created by the perception of time is also closed. This quanta of information is analogous to our graviton, and the basis for developing a theory of quantum gravity. The reason why gravitons have been impossible to detect by quantum physicists thus far probably has to do with the fact that they are so deeply embedded in the electromagnetic force, and that they have no spin, nor any angular momentum. However, they too transform the oscillating frequency of a wave into the vibratory frequency of a particle in a controlled space. Some waves moved upon the face of the deep space warps at angle, or diagonally across an imaginary line laid across the face thereof. These are the waves that entered into a spin to behave like a particle drawing the particle out of infinity. This angular entry into these space warps is the primeval basis for the theories of angular momentum known today. Remember, gravitons are deeply embedded in the electromagnetic force. Therefore, to fully understand the graviton, we such have basic understanding of an electromagnetic wave (or electromagnetic radiation) with regard to it moving along a linear line. Electromagnetic radiation is a combination of oscillating electric and magnetic fields moving through a medium perpendicular to each other through space and carrying energy from one place to another. Therefore, when a an electromagnetic wave passed across a space warps parallel to the imaginary line laid across it, the perpendicular electromagnetic field collapses into the wave itself to form a graviton. The collapse to the electromagnetic field can be interpreted as the absorption of quanta of energy by the electromagnetic wave, or the graviton itself. Thus the value of the absorption of quanta of energy can be calculated from the point of origin to any point down the perpendicular line as it descends down the tiny black hole—the point origin being the point at which the electromagnetic field began to collapse into the electromagnetic wave. Therefore, a graviton is an accumulative value. That is the basis for developing a theory of quantum gravity and reconciling it to the geometrical description of gravity in Einstein's general relativity. Therein, it should be made crystal clear how both gravitons and gravity analogously employed the other three fundamental forces of nature to create the physical and natural Universe as it is now seen. When we consider the fact that a graviton is an accumulative value, analogous to the accumulation of particles or atoms to produce a mass heavy enough to exert a gravitational pull on smaller mass, the reconciliation shouldn't be too difficult. There is something else we should consider about the graviton and its relationship with a space warp. It has long been theorized that a space warp also produces a time warp that overcomes the limitations of the speed of light. The existence of such a phenomenon would be highly consistent with the Will of God to overcome the perception of time and space in order to be Omnipresent and Omnibus. Remember, electromagnetic waves travel at the speed of light. Now let us consider the fact that a graviton is formed by electromagnetic waves being drawn to a point of singularity from opposite ends of infinity at the same time. Thus we can be relatively certain that the effective speed of the movement of this energy along a linear line is some multiple of the of the speed of light—maybe even as fast as the speed of light squared. I'm sure some of the great mathematicians will figure it out. Meanwhile, there is something else we can't be too sure at this point. We can't be sure if this energy is stored in atoms or the graviton itself as potential energy, of around the graviton and/or its vibratory frequency as kinetic energy. Now that they know where to look and what to look for, I'm sure some of the great quantum physicists of our day will figure that out also. Finally we must remember Planck epoch, the time when the universe was sufficiently dense and energetic that all the fundamental forces, including gravity, were merged into one grand force. Well now we can be relatively certain that the graviton was the catalyst for the great merger of the fundamental forces of nature...

References

Amen, Nur Ankh. 2001. *The Ankh: African Origin of Electromagnetism*. Buffalo, NY: Eworld, Inc.

Ashby, Muata Abhaya. 2005. *Introduction to Maat Philosophy*. Miami, FL: Sema Institute.

Ashby, Muata Abhaya. 2005. *Egyptian Yoga II: The Supreme Wisdom of Enlightenment*. Miami, FL: Sema Institute.

Ashby, Muata Abhaya. 2005. *The Kemetic Diet: Food For Body, Mind, and Soul, A Holistic Health Guide Based on Egyptian Medical Teachings*. Miami, FL: Sema Institute.

Akbar, Na'im. 1998. *Know Thyself*. Tallahassee, FL: Mind Productions & Associates.

Aurobindo, Sri. 1968. *Sri Aurobindo or The Adventure of Consciousness*. India: Sri Aurobindo Ashram.

Avalon, Arthur. 1974. *The Serpent Power: The Secrets of Tantric and Shaktic Yoga*. Mineola, NY: Dover Publications.

Ballentine, Rudolph Rama Swami, and Hymes, Alan. 1979. *Science of Breath: A Practical Guide*. Honesdale, PA: The Himalayan Institute Press.

Barnes, Carol. 2001. *Melanin: The Chemical Key to Black Greatness*. Bensenville, IL: Lushena Books.

Beinfield, Harriet, and Korngold, Efrem. 1991. *Between Heaven and Earth: A Guide to Chinese Medicine.* New York, NY: Ballantine Books.

Beye, Taj Tarik. 2009. *"Sovereignty, UCC, Excise Tax and History."* Detroit, MI.

Browder, Anthony T. 1996. *From the Browder File: Survival Strategies for Africans in America 13 steps to Freedom.* Washington, DC: Inst. Of Karmic Guidance.

Browder, Anthony T. 1992. *Nile Valley Contributions to Civilization: Exploding the Myths Vol. I* Washington, DC: Inst. Of Karmic Guidance.

Burger, Bruce. 1998. *Esoteric Anatomy.* Berkeley, CA: North Atlantic Books

Capra, Fritjof. 1975. *The Tao of Physics: An Exploration of the Parallels between Modern Physics and Eastern Mysticism.* Boston, MA: Shambhala Publications.

Chen, John K. 2004. *Chinese Medical Herbology and Pharmacology.* City of Industry, CA: Art of Medicine Press.

Cheung, William. 1986. *How to Develop Chi Power.* New York, NY: Black Belt Books.

Chia, Montak. 2006. *Bone Marrow Nei Kung: Taoist Techniques for Rejuvenating the Blood and Bone.* Rochester, VT: Destiny Books.

Chia, Montak. 2007. *Fusion of the Five Elements: Meditations for Transforming Negative Emotions.*

Rochester, VT: Destiny Books.

Chia, Montak. 2007. *Chi Nei Tsang: Chi Massage for the Vital Organs.* Rochester, VT: Destiny Books.

Chia, Montak. 2009. *The Six Healing Sounds: Taoist Techniques for Balancing Chi.* Rochester, VT: Destiny Books

Chia, Montak. 2008. *The Inner Smile: Increasing Chi through the Cultivation of Joy.* Rochester, VT: Destiny Books

Clarke, John Henrik. 2002. *Christopher Columbus and the Afrikan Holocaust: Slavery and the Rise of European Capitalism.* Brooklyn, NY: A&B Distributors.

Cleary, Thomas. 2000. *Taoist Meditation: Methods for Cultivating a Healthy Mind and Body.* Boston, MA: Shambhala.

Diop, Cheikh Anta. 1989. *The African Origin of Civilization: Myth or Reality.* Chicago, IL: Lawrence Hill Books.

Diop, Cheikh Anta. 1988. *Precolonial Black Africa.* Chicago, IL: Lawrence Hill Books.

Diop, Cheikh Anta. 1991. *Civilization or Barbarism: An Authentic Anthropology.* Chicago, IL: Lawrence Hill Books.

Duff, Robert M. 2006. *A Different Future: An Introduction to an Esoteric Form of Raja Yoga.*

Einstein, Albert. 1961. *Relativity: The Special and General Theory.* New York, NY: THREE RIVERS PRESS.

Elliot, W. Scott. 1925. *Legends of Atlantis and Lost Lumeria*. Chennai, India: Quest Books.

Epperson, Ralph A. 1990. *New World Order*. Chattanooa, TN: Publius Press.

Gadalla, Moustafa. *Exiled Egyptians: The Heart of Africa*. Grensboro, NC: Tehuti Research Foundation.

Gadalla, Moustafa, *Historical Deception: The Untold Story of Ancient Egypt*. Greensboro, NC: Tehuti Research Foundation.

Garner, Bryan A., and Schultz, David W. 1999. *A Handbook of Basic Law Terms*. Farmington Hills, MI: Gale Cengage.

Garner, Bryan A. 2006. *Black's Law Dictionary*. Eagan, MN: Thomson West.

Gray, Henry F.R.S. 1977. *Gray's Anatomy*. New York, NY: Crown Publishers, Inc.

Gremillion, Zachary P. 2005. *African Origins of Freemasonry: Treatise of the Ancient Grand Lodge Khamet*. Bloomington, IN: AuthorHouse.

Hawking, Stephen. 2001. *The Universe in a Nutshell*. New York, NY: Bantam Books.

Hoehn, Katja, and Marieb, Elaine. 2008. *Anatomy & Physiology*. Upper Saddle, NJ: Benjamin Cummings.

Howell, Edward Dr. 1985. *Enzyme Nutrition: The Food Enzyme Concept*. New York, NY: Penguin Putnam, Inc.

Hseuh, Chen Chiu. 1981. *Acupuncture: A Comprehensive Text.* Seattle, WA: Eastland Press.

Jwing-Ming, Yang. 2003. *Qigong Meditation: Embryonic Breathing.* Wolfeboro, NH: YMAA Publication Center.

Jwing-Ming, Yang. 2006. *Qigong Meditation: Small Circulation.* Wolfeboro, NH: YMAA Publication Center.

Jwing-Ming, Yang. 1997. *The Root of Chinese Qigong: Secrets of Health, Longevity, & Enlightenment.* Wolfeboro, NH: YMAA Publication Center.

Jwing-Ming, Yang. 2000. *Qigong, The Secret of Youth: Da Mo's Muscle/Tendon Changing and Marrow/Brain Washing Classics.* Wolfeboro, NH: YMAA Publication Center.

Judith, Anodea. 1994. *Wheels of Life: A User's Guide to the Chakra System.* St. Paul, MN: Llewellyn Publications.

Kalat, James W. 2006. *Biological Psychology.* Stamford, CT: Wadsworth Publishing.

Krishnamurti, J. 1973. *The Awakening of Intelligence.* New York, NY: Harper San Francisco

Krishnamurti, J. 1975. *Freedom From the Known.* New York, NY: Harper San Francisco

Krishnamurti, J. 1971. *At the Feet of the Master.* Adyar, India: Theosophical Pub House.

Krishnamurti, J. 1989. *Think on These Things.* New York, NY: HarperOne.

Lu, Henry C. 2005. *Traditional Chinese Medicine: An Authoritative and Comprehensive Guide.* New York, NY: Basic Health Publications.

Lu, Henry C. 2006. *Chinese Natural Cures: Traditional Methods for Remedy and Prevention.* New York, NY: Black Dog & Leventhal Publishers.

Lurker, Manfred. *The Gods and Symbols of Ancient Egypt: An Illustrated Dictionary.* London, England: Thames & Hudson.

Maciocia, Giovanni. 2007. *The Practice of Chinese Medicine: The Treatment of Diseases with Acupuncture and Chinese Herbs.* New York, NY: Churchill Livingstone.

Maciocia, Giovanni. 2005. *The Foundations of Chinese Medicine: A Comprehensive Text for Acupuncturists and Herbalists.* New York, NY: Churchill Livingstone.

Mann, Felix. 1972. *Acupuncture: The Ancient Chinese art of Healing.* New York, NY: Random House, Inc.

Moore, John T. 1955. *Chemistry Made Simple: A Complete Introduction to the Basic Building Blocks of Matter.* New York, NY: Random House, Inc.

Muhammad, Dr. Ali. 2008. *"Moorish Origin of Civilization."* Detroit, MI.

Muhammad, Dr. Ali. 2008. *"The Melanated Body and Diet."* Detroit, MI.

Myers, Timothy. 2005. *The Huevolution of Sacred Muur Science Past and Present: A Theoretical Compilation.*

Bloomington, IN: AuthorHouse.

Parker, George Wells. 1978. *Children of the Sun.* Clattern House, Kingston, Surrey: BCB.

Paulson, Genevieve L. 2002. *Kundalini & the Chakras: Evolution in this Lifetime.* Woodbury, MN: Llewellyn Publications.

Pitchford, Paul. 2002. *Healing With Whole Foods: Asian Traditions and Modern Nutrition.* Berkeley, CA: North Atlantic Books

Pookrum, Jewel. 2000. *Vitamins & Minerals from A to Z with Ethno-Consciousness.* Brooklyn, NY: A & B Distributors.

Ra Un Nefer Amen. 1990. *Metu Nter Vol. I: The Oracle of Tehuti and the Egyptian System of spiritual Cultivation.* Philadelphia, PA: Khamit Medi Trans Visions, Inc.

Reid, Daniel P. 1989. *The Tao of Health, Sex, and Longevity: A Modern Practical guide to the Ancient Way.* New York, NY: Fireside Books.

Richens, David T. 1997. *The Chemistry of Aqua Ions: Synthesis, Structure and Reactivity: A Tour Through the Periodic Table of the Elements.* Hoboken, NJ: Wiley.

Roach, Geshe Michael. 2005. *The Essential Yoga Sutra: Ancient Wisdom for Your Yoga.* New York, NY: Three Leaves.

Rogers, J.A. 1985. *100 Amazing Facts About the Negro, with Complete Proof: A Short Cut to the World History of the Negro.* Saint Petersburg, FL: Helga M. Rogers.

Samuels, Mike, and Samuels, Nancy. 1975. *Seeing With The Mind's Eye: The History, Techniques and Uses of Visualization.* New York, NY:Random House, Inc.

Sebi, Dr.2009. "*Electric Food: Holistic Health and Nutrition.*" Detroit, MI.

Sedona, Gabriel. 1995. *The Cosmic Family Volume I.* New York, NY: Extension Schools of Melchizedek.

Sertima, Ivan Van. 2003. *They Came Before Columbus: The African Presence in Ancient America.* New York, NY: Random House Trade.

Sertima, Ivan Van. 1991. *The Golden Age of the Moor.* Newark, NJ: Transaction Publishers.

Sertima, Ivan Van. 1986. *African Presence in Early Europe.* Newark, NJ: Transaction Publishers.

Sertima, Ivan Van. 1987. *African Presence in Early Asia.* Newark, NJ: Transaction Publishers.

Sertima, Ivan Van. 1987. *African Presence in Early America.* Newark, NJ: Transaction Publishers.

Shou-Yu Liand, and Wen-Ching Wu. 1997. *Qigong Empowerment: A Guide to Medical Taoist Buddhist Wushu Energy Cultivation.* East Providence, RI: The Way of Dragon Publishing.

Sills, Franklyn. 2001. *The Polarity Process: Energy as a Healing Art.* Berkeley, CA: North Atlantic Books.

Sitchin, Zecharia. 1990. *Genesis Revisited.* New York, NY: Avon.

Sitchin, Zecharia. 2004. *The Lost Book of Enki: Memoirs and Prophecies of an Extraterrestrial God.* Rochester, VT: Bear & Company.

Sitchin, Zecharia. 2007. *Twelfth Planet: Book I of the Earth Chronicles.* New York, NY: Harper.

Sitchin, Zecharia. 2007. *The Stairway to Heaven: Book II of the Earth Chronicles.* New York, NY: Harper.

Sitchin, Zecharia. 2007. *The Wars of Gods and Men: Book III of the Earth Chronicles.* New York, NY: Harper.

Sitchin, Zecharia. 2007. *When Time Began: Book V of the Earth Chronicles.* New York, NY: Harper

Trungpa, Chogyam. 2001. *Crazy Wisdom.* Boston, MA: Shambahla.

Trungpa, Chogyam. 2003. *Training the Mind and Cultivating Loving Kindness.* Boston, MA: Shambhala.

Trungpa, Chogyam. 2007. *Shambhala: The Sacred Path of the Warrior.* Boston, MA: Shambhala.

Trungpa, Chogyam. 2001. *Great Eastern Sun; The Wisdom of Shambhala.* Boston, MA: Shambhala.

Trungpa, Chogyam. 1996. *Meditation in Action.* Boston, MA: Shambhala.

Tsu, Lao, and Feng, Gia-Fu. 1989. *Tao Te Ching*. New York, NY: Vintage.

Vivekananda, Swami. 2007. *The Yoga Sutras of Patanjali: The Essential Yoga Text for Spiritual Enlightenment*. London, UK: Watkins.

Vivekananda, Swami. 1982. *Jnana Yoga*. New York, NY: Ramakrishna-Vivekananda Center.

Vivekananda, Swami 1982. *Karma-Yoga and Bhakti-Yoga*. New York, NY: Ramakrishna-Vivekananda Center

Vivekananda, Swami. 1980. *Raja-Yoga*. New York, NY: Ramakrishna-Vivekananda Center.

Welsing, Frances Cress. 1991. *The Isis Papers: The Keys to the Colors*. Chicago, Il: Third World Press.

Wilhelm, Richard, and Baynes, Cary F. 1950. *The I Ching or Book of Changes*. New York, NY: Bollingen Foundation, Inc.

Williams, Chancellor. 1987. *Destruction of Black Civilization: Great Issues of a Race from 4500 B.C. To 2000 A.D.* Chicago, IL: Third World Press.

Yogananda, Paramahansa. 1937. *Man's Eternal Quest*. Los Angeles, CA: Self-Realization Fellowship.

Yogananda, Paramahansa. 2001. *God Talks with Arjuna: The Bhagavad Gita*. Los Angeles, CA: Self-Realization Fellowship.

York, Malachi. 1994. *The Sacred Wisdom: Hierophant Tehuti (Forward Through the Past)* Holy Tabernacle Ministries.

Eatonton, GA.

York, Malachi. 1996. *The Holy Tablets.* Holy Tabernacle Ministries. Eatonton, GA.

York, Malachi. 2008. *Yashua Found in Egipt.* TamaRe House. Eatonton, GA.

York, Malachi. 1994. *El's Holy Injiyl.* Holy Tabernacle Ministries. Eatonton, GA.

York, Malachi. 1985. *El's Holy Qur'aan.* Holy Tabernacle Ministries. Eatonton, GA.

York, Malachi. 1988. *The Book of Light.* Holy Tabernacle Ministries. Eatonton, GA.

York, Malachi. 1990. *The Sacred Records of Neter.* Holy Tabernacle Ministries. Eatonton, GA.

York, Malachi. 1993. *El's Holy Tehillim(Psalms).* Holy Tabernacle Ministries. Eatonton, GA.

York, Malachi. 1995. *El Katub Shil El Mawut(The Book of the Dead).* Holy Tabernacle Ministries. Eatonton, GA.

www.ingramcontent.com/pod-product-compliance
Lightning Source LLC
Chambersburg PA
CBHW031949080426
42735CB00007B/319